The Law and Practice Relating to Pollution Control in Denmark

There are nine other titles in this series:

The Law and Practice Relating to Pollution Control In

 Belgium and Luxembourg
 France
 Federal Republic of Germany
 Greece
 Ireland
 Italy
 The Netherlands
 The United Kingdom

The Law and Practice Relating to Pollution Control in the Member States of the European Communities: A Comparative Survey

The series will be updated at regular intervals. For further information, complete the enclosed postcard and send it to:
Graham & Trotman Limited
Sterling House
66 Wilton Road
London SW1V 1DE

All the titles in the series were prepared by

Environmental Resources Limited
79 Baker St, London W1M 1AJ (Tel. 01-486 8277; Tx. 296359 ERL G)

for

The Commission of the European Communities,
Directorate-General Environment, Consumer Protection
and Nuclear Safety, Brussels

The Law and Practice Relating to Pollution Control in Denmark

Second Edition

Prepared by

Professor C. H. Jensen
Aalborg University

for

Environmental Resources Limited

Published by
Graham & Trotman
for
The Commission of the European Communities

Published in 1982 by

Graham & Trotman Limited
Sterling House
66 Wilton Road
London SW1V 1DE

for

The Commission of the European Communities,
Directorate-General Information Market and Innovation,
Luxembourg

EUR 7733

© ECSC, EEC, EAEC, Brussels and Luxembourg, 1982

British Library Cataloguing in Publication Data

Jensen, C. H.
   The law and practice relating to pollution
   control in Denmark.—2nd ed.
   1. Pollution—Law and legislation—Denmark
   I. Title   II. Environmental Resources Ltd.
   III. Commission of the European Communities
   344.8904'463

ISBN 0-86010-312-9

The views expressed in this publication are those of the author, and should not be taken as reflecting the opinion of the Commission of the European Communities.

LEGAL NOTICE

Neither the Commission of the European Communities nor any person acting on behalf of the Commission is responsible for the use which might be made of the following information.

All rights reserved. No part of this publication may be reproduced, stored in a retrieval system, or transmitted in any form or by any means, electronic, mechanical, photocopying, recording or otherwise, without the prior permission of the publishers.

Printed in Great Britain by
Robert Hartnoll Limited, Bodmin, Cornwall

Summary List of Contents

1	Authorities and Pollution Control	1
2	Air	47
3	Fresh Water	80
4	Pollution of the Sea by Substances other than Oil	105
5	Pollution of the Sea by Oil	119
6	Discharge into the Sewerage System	132
7	Waste Disposal on Land	141
8	Noise and Vibration	155
9	Radioactivity	174
10	Product Controls	188
11	Environmental Impact Assessment	203
12	Closing Remarks	207

Preface

This volume is part of a series prepared in the performance of a contract between the Commission of the European Communities and Environmental Resources Limited (ERL). ERL is a consulting organisation specialising in environmental research, planning and management.

In 1976 a first series was published covering the, then, nine members of the Community. The purpose of those volumes was to explain the law and practice of pollution control in each of the Member States and to provide a summary comparing all the countries in a separate comparative volume.

Since that time many changes in legislation arising from both national and Community-wide initiatives have occurred. ERL was therefore asked to prepare a new series providing an up-to-date review of the law and practice relating to pollution control in the Member States of the European Community.

The series comprises nine volumes concerning the law and practice in the Member States:

Belgium and Luxembourg
Denmark
France
The Federal Republic of Germany
Greece
Ireland
Italy
The Netherlands
The United Kingdom

and a summary comparative volume.

The aim of this new series, as in the first, is to provide a concise but fully referenced summary of the letter of the law, and a discussion of its implementation and enforcement in practice. Proposals for new legislation which has been drafted but not yet passed are outlined. Where laws have been introduced to comply with Community-wide requirements this is noted.

PREFACE

The publication has two principal objectives:

> to enable the reader to study in outline the provisions in any one Member State; and

> to enable a direct comparison between different Member States.

To facilitate comparison between the national reports, each is indexed following a standard format (the Classified Index) to enable easy reference to the relevant sections of each report.

Presenting a nation's laws accurately in summary form is always a difficult task. There is a danger that, out of context, they may be misunderstood. We have therefore tried to give, in the first section of each report, some of the constitutional, legal and administrative background.

A further danger lies in translation. Although in the English texts we have tried to prepare as accurate a translation as possible, only the authors' original texts in their native languages carry their full authority. These texts are also being published in the individual Member States.

The statement of law in each volume is correct to at least 30 June 1981; in some cases more recent revisions have been included during the period of preparation for publication.

The series will be updated at regular intervals; to receive further details readers should complete the enclosed postcard and send it to the publisher.

ERL would like to acknowledge and express its thanks for the contributions from the national authors and for their cooperation in the preparation of the series.

Finally, ERL also acknowledges the assistance provided by many agencies, which have freely given information and advice, and the help and guidance given by Monsieur Claude Pleinevaux, Mr Grant Lawrence and other members of the Directorate of Environment, Consumer Protection and Nuclear Safety of the Commission of the European Communities.

1982   Environmental Resources Limited
London

Detailed List of Contents

Summary List of Contents	v
Preface	vii

1 Authorities and Pollution Control — 1

1.1 Constitution and Public Administration — 1
   1.1.1 Some Characteristics of the Danish Constitution — 1
   1.1.2 The General Structure of Administration — 2

1.2 Sources of Law in the Field of Environmental Protection — 6

1.3 Central Authorities — 9

1.4 Regional and Local Authorities — 14

1.5 Independent Consultative Bodies — 20

1.6 Rights of Special Interest Groups — 20
   1.6.1 Representation on Administrative Bodies — 21
   1.6.2 The Right to be Consulted — 22
   1.6.3 Other Rights — 23

1.7 Rights of the Individual — 24
   1.7.1 The Right to Initiate Administrative Proceedings — 25
   1.7.2 The Right to be Heard, etc. — 25
   1.7.3 The Right of Appeal — 28
      1.7.3.1 Individuals with the right of appeal other than the addressee — 31
      1.7.3.2 Associations entitled to appeal — 33
      1.7.3.3 Notice of the decision of the lower administrative authority — 34
   1.7.4 Suspensive Effect of Appeals — 35
   1.7.5 The Right of Action *(Locus Standi)* — 36
   1.7.6 Actions for Compensation — 38
   1.7.7 The General Law of Adjoining Properties — 41

CONTENTS

2 Air   47

2.1 Stationary Sources of Air Pollution   47
   2.1.1 Control by the General Planning Legislation   47
   2.1.2 Specific Environmental Control as Regards the Location, Arrangement and Operation of Plants   53
      2.1.2.1 The system of approvals under the Environmental Protection Act   53
      2.1.2.2 Specific orders and prohibitions under the Environmental Protection Act   57
      2.1.2.3 General rules   59
   2.1.3 Requirements as to Purifying Processes, etc. Prior to the Discharge of Pollutants into the Air   60
      2.1.3.1 Arrangements for purification   60
      2.1.3.2 Chimney height, etc.   61
   2.1.4 Emission Limits   63
   2.1.5 Duty of the Polluter to Monitor his Emissions into the Air   64
   2.1.6 Enforcement   65
      2.1.6.1 Monitoring   65
      2.1.6.2 Sanctions   66
      2.1.6.3 Suspension of enforcement   68
   2.1.7 Air Quality Objectives   68
      2.1.7.1 Legal requirements and recommendations   68
      2.1.7.2 Details concerning the observance of air quality standards   69
   2.1.8 Rights of the Individual   69
      2.1.8.1 Right to information   69
      2.1.8.2 Right to appeal against permits   70
      2.1.8.3 Right to make statements   70
      2.1.8.4 Right to raise or to intervene in environmental cases   71
      2.1.8.5 *Locus standi* (right to take legal action)   71

2.2 Motor Vehicles   71
   2.2.1 Design and Equipment   71
   2.2.2 Maintenance   72
   2.2.3 Use   73
   2.2.4 Requirements Concerning the Fuel Used in Motor Vehicles   73
   2.2.5 Enforcement   74
   2.2.6 Rights of Individuals   75

2.3 Aeroplanes, Hovercraft and Ships   75
   2.3.1 Aeroplanes   75
   2.3.2 Hovercraft   76
   2.3.3 Ships   76

CONTENTS

3 Fresh Water   80

3.1 Stationary Sources of Water Pollution   81
    3.1.1 Control through General Planning Legislation   81
    3.1.2 Preventive Control   81
        3.1.2.1 Protection areas in connection with water supply installations   82
        3.1.2.2 Subterranean containers for oil, etc.   82
        3.1.2.3 Other preventive controls   84
    3.1.3 Pretreatment   85
    3.1.4 Control over the Composition and Amounts, etc. of Effluent Discharged   85
        3.1.4.1 Permit schemes   86
            3.1.4.1.1 Conveyance to surface water   86
            3.1.4.1.2 Discharge into the ground   87
            3.1.4.1.3 Situation with regard to Chapter 5 of the Environmental Protection Act   88
        3.1.4.2 Intervention in existing discharge arrangements, etc.   88
        3.1.4.3 Material criteria for decisions concerning discharge   90
            3.1.4.3.1 Protection of water recipients   91
            3.1.4.3.2 Protection of soil recipients   94
        3.1.4.4 Charges for waste-water disposal   94
    3.1.5 Control of Dumping in Fresh Water or on Nearby Land   95
    3.1.6 Supervision on the Part of the Polluter   95
    3.1.7 Enforcement   95
    3.1.8 Water Quality Targets   97
        3.1.8.1 Legal requirements and recommended guidelines   97
        3.1.8.2 Information on observance of the water quality targets   98
        3.1.8.3 Restrictions on the use of water which does not meet the quality requirements   98
    3.1.9 Individuals' Rights   99

3.2 Pollution of Fresh Water by Ships   101
    3.2.1 Shipbuilding and Fitting Out   101
    3.2.2 Maintenance   101
    3.2.3 Use   101

4 Pollution of the Sea by Substances other than Oil   105

4.1 Coastal Waters   106
    4.1.1 Definition of Limits   106
    4.1.2 Control by Means of Ordinary Planning Legislation   107
    4.1.3 Preventive Controls   108
    4.1.4 Pretreatment   108
    4.1.5 Control of the Composition and Amount, etc. of Discharges   108
    4.1.6 Supervision on the Polluter's Part   109
    4.1.7 Enforcement   109

CONTENTS

4.1.8 Quality Targets for Coastal Waters	110
4.1.8.1 Legal requirements and recommended guidelines	110
4.1.8.2 Information concerning observance of the targets	111
4.1.8.3 Restrictions on the Use of Seawater that Does Not Meet the Targets	111
4.1.9 Individuals' Rights	111
4.2 Control of Dumping from Ships and Aircraft	**112**
4.2.1 Geographical Extent of Danish Jurisdiction and Substantive Rules	113
4.2.2 Supervision	114
4.2.3 Enforcement	115
4.3 Control of Pollution from Exploitation of the Sea or the Sea Bed	**115**
4.3.1 Geographical Extent of Danish Jurisdiction and Substantive Rules	115
4.3.2 Supervision	116
4.3.3 Enforcement	117

5 Pollution of the Sea by Oil   119

5.1 Control of Discharge from Ships	**119**
5.1.1 The Fitting Out and Equipment of a Ship	119
5.1.2 Ship's Crew	120
5.1.3 Emptying of Oil	120
5.1.4 Installations for Receiving Oil Residues	121
5.1.5 Enforcement	122
5.1.5.1 Supervision	122
5.1.5.2 Countermeasures	123
5.1.6 Civil Law Liability in Damages	124
5.2 Control over Installations on the Coast and Activities in Ports	**126**
5.2.1 Siting and Construction of Installations on the Coast	126
5.2.2 Loading and Unloading	126
5.2.3 Enforcement	126
5.3 Control over Fixed Installations at Sea	**127**
5.3.1 Extent of Danish Jurisdiction	127
5.3.2 Approval Schemes	127
5.3.3 Construction, Equipment and Safety Zones	127
5.3.4 Manning	128
5.3.5 Discharge of Oil	128
5.3.6 Loading and Unloading	129
5.4 Contingency Plans for Dealing with Oil Pollution Accidents	**129**

CONTENTS

6 Discharge into the Sewerage System — 132

   6.1 Bans and Permits — 133

   6.2 Control over Composition and Quantities, etc. for Effluent Discharged into the Sewerage System — 135
      6.2.1 Composition and Quantities — 135
      6.2.2 Charges — 136

   6.3 Requirements for the Purification of Waste-water — 137

   6.4 Control of Discharge from Sewage and Purification Plants — 139

   6.5 Enforcement — 139

   6.6 Civil Law Liability in Damages — 139

7 Waste Disposal on Land — 141

   7.1 Control of the Siting of Installations for the Deposit and Treatment of Waste — 142

   7.2 Requirements Concerning Treatment Prior to Deposit of the Waste, with a View to Recycling — 143

   7.3 Control of the Method of Disposal — 144
      7.3.1 Compulsory Removal Schemes — 145
      7.3.2 Treatment or Storage of Waste after Removal — 146
      7.3.3 Establishment and Operation of Installations for the Deposit or Treatment of Waste — 146

   7.4 Restoration of the Ground after Tipping — 147

   7.5 Enforcement — 147

   7.6 Controls of Special Forms of Waste — 148
      7.6.1 Waste Oil — 148
      7.6.2 Chemical Waste — 150
      7.6.3 Other Special Forms of Waste — 150

   7.7 Control of Special Categories of Products — 151

   7.8 Individuals' Rights — 152

8 Noise and Vibration — 155

   8.1 Stationary Sources — 155

        8.1.1 Control over the Siting of Installations which Produce Noise and Vibration   155
        8.1.2 Control over the Establishment and Use of Noisy Installations and Equipment   159
        8.1.3 Emission Limits   160
        8.1.4 Enforcement   163
        8.1.5 Noise Quality Targets   163
            8.1.5.1 Legal requirements and recommended guidelines   163
            8.1.5.2 Details of the observance of quality standards   164
            8.1.5.3 Restriction of activities in noise-ridden areas   164
            8.1.5.4 Insulation requirement   166
        8.1.6 Individuals' Rights   166

  8.2 Motor Vehicles   167
        8.2.1 Construction and Equipment   167
        8.2.2 Emission Standards   167
        8.2.3 Maintenance   168
        8.2.4 Use   169
        8.2.5 Enforcement   169
        8.2.6 Individuals' Rights   170

  8.3 Aircraft   170
        8.3.1 Construction and Equipment   170
        8.3.2 Emission Standards   170
        8.3.3 Maintenance   170
        8.3.4 Use   171
        8.3.5 Enforcement   171
        8.3.6 Individuals' Rights   172

  8.4 Ships, Including Hovercraft   172

9 Radioactivity   174

  9.1 Nuclear Plants   174
        9.1.1 Control of the Siting of Atomic Plants   176
        9.1.2 Control of the Construction of Atomic Installations   177
        9.1.3 Control of Operation and Maintenance   177
        9.1.4 Control of Nuclear Waste   179
        9.1.5 Supervision by the Polluter   179
        9.1.6 Enforcement   179
        9.1.7 Individuals' Rights   180
            9.1.7.1 The right to obtain information   180
            9.1.7.2 Right of appeal   181
            9.1.7.3 Right to have views taken into consideration   181
            9.1.7.4 Right to institute and intervene in enforcement proceedings   182
            9.1.7.5 Right to compensation   182

9.2	Radioactive Substances	183
	9.2.1 Control over Siting	183
	9.2.2 Control of Storage and Use	183
	9.2.3 Control of Packaging and Transport	185
	9.2.4 Control of Disposal	185
	9.2.5 Supervision on the Polluter's Part	185
	9.2.6 Quality Targets	185
	9.2.7 Enforcement	186
	9.2.8 Individuals' Rights	186

10 Product Controls — 188

10.1 General Aspects of the Chemicals Act — 189

10.2 Dangerous Chemical Substances and Products — 195

10.3 Pesticides — 197

10.4 Use of Poisons and Substances Dangerous to Health for Special Purposes — 199

10.5 Other Regulations Pursuant to the Chemicals Legislation — 200

10.6 Lead Shot — 201

10.7 Oil Products — 201

10.8 Packaging — 201

10.9 Radioactive Substances — 201

11 Environmental Impact Assessment — 203

12 Closing Remarks — 207

Classified Index — 213

1
Authorities and Pollution Control

1.1 CONSTITUTION AND PUBLIC ADMINISTRATION

1.1.1 Some characteristics of the Danish Constitution

In an account which forms part of a comparative study I think there is reason to start by giving a brief description of the Danish Constitution. It goes without saying that such an introduction must be limited to a few elements in outline.

Legally speaking, Danish government is based first and foremost on a written constitution, the latest version being the Constitution of the Kingdom of Denmark of 5 June 1953. This Constitution forms a very stable foundation for the administration as the procedure for its amendment is comprehensive and onerous. Indeed, under section 88 of the Danish Constitution, a proposed amendment must first be adopted by two sessions of the Folketing (Danish parliament) separated by a general election, and then approved in a referendum in which a majority of those taking part and at least 40% of those entitled to vote are in favour of the proposed amendment. These requirements for effecting amendments to the Constitution are considerably more stringent than the rules for effecting amendments to the constitutions of most other countries in Western Europe.

The basic characteristics of the Danish Constitution may be summed up roughly in four points:

(1) The mode of government is intrinsically by parliamentary democracy. This implies firstly that the legislative power is in the hands

of a parliament elected by the people—the Folketing—and the Government and, secondly, that the Government is subject to the parliamentary system both as regards its legislative and executive powers, i.e. the Government must resign if a majority of members of parliament are against it.[1]

(2) A number of fundamental freedoms or human rights are protected by various provisions in the Constitution. The rights in question are, in particular, personal liberty, property rights, freedom of expression, freedom of association and freedom of assembly.[2]

(3) The Constitution provides for the establishment of a system of local government.[3] The system is to be implemented in detail by statute, however, and it is specifically laid down that local government shall be exercised under central government supervision. Thus the Danish Constitution does not provide for the setting up of a federal state in the proper sense of the term.[4]

(4) The public administration—central and local alike—is subject to legality control by the courts of law.[5] This control is exercised by the ordinary courts. It is compatible with the Constitution to establish special administrative courts, but this is rendered impractical by a constitutional requirement to the effect that the decisions of such tribunals may be brought before the Supreme Court and there are at the moment no administrative courts.

1.1.2 The general structure of administration

For a proper understanding of the background to the administrative organisation in the field of environmental protection described below in 1.3.5, it will be necessary first to give an outline of the structure of the Danish administrative machinery. In accordance with recent expositions of administrative law, such a survey of the administrative organisation may best be provided by describing the most important types of administrative authorities.[6]

Before dealing with the individual types of administrative authority, I would like to establish that by tradition Denmark has a *comparatively centralised administration*. This is partly because the Constitution provides a basis for the administrative tasks to be carried out by ministries and authorities subordinate to the ministries. A bond is thereby established between the administration and the legislature as ministers who are politically responsible to parliament (Folketing) are in charge of the ministries and a considerable measure of coordination of administrative work is ensured. The Constitution does not, however, contain any provisions as to the detailed structure of the administrative machinery. The structure is determined partly by statute, partly by the ministers whose decisions on organisation are subject to approval by the Folketing if financial appropriations are necessary. Many or few tasks may be assigned to the ministry or to subordinate authorities and it is possible within rather wide limits to entrust administrative functions to bodies enjoying considerable independence, especially boards and local councils. This freedom to assign administrative tasks has been increasingly utilised by the legislature in recent years. This has led to increasing *decentralisation*, but it should be noted that the powers of the ministries have in several respects been extended compared with what they used to be. Therefore the most striking feature in the development may be said to be a change from detailed control to a system of overall control.

As already mentioned, the *ministries* are headed by the respective politically appointed ministers. The personnel of a ministry is hierarchically organised under the minister and consists of staff with qualifications in law, economics, political science or other training which fits them for administrative work. The ministries may be large or small depending partly on the field of work, partly on the existence of subordinate authorities within the respective field. The most important tasks of a ministry are typically the preparation of Bills, the issuing of general directives and instructions, the planning of administrative participation and the administration of appropriations. Decisions on some important individual cases fall within the province of some ministries, e.g. petitions for exemption and complaints about decisions made by subordinate authorities. In recent years, however, in an effort to improve the ministries' capacity to exercise their directive-issuing and planning functions, the legislature has been cutting down in more and more sectors on access to the ministries to lodge complaints about administrative decisions on concrete cases.[8] In such circumstances concrete cases will then usually be referred to a directorate or a board.

The closest connection exists between the ministries and the *directorates*. The latter are also central authorities, hierarchically organised, but they differ from the ministries in several respects. The head of a directorate, the director, is a civil servant, not a politician, and some members of the staff will typically be specialists with, for example, various kinds of technical training. The sphere of activity of a directorate is usually narrower than that of a ministry, so that often several directorates come under the same ministry, the most important tasks of a directorate usually being to decide a great number of individual cases and to supervise local administrative bodies. The ministry is usually superior in all respects to the directorate. It may request information from the directorate on its own initiative or, following a complaint, it may reverse decisions made by a directorate and issue directives concerning the functions of the latter. A characteristic trend in the 1960s and more particularly in the 1970s was that, in order to relieve the ministries, a considerable number of new directorates were established and the functions of the directorates were expanded compared with the previous situation.[9] This substantial enlargement of the directorates at the expense of the ministries probably involved a tendency towards the limitation in practice of the ministries' legal powers to control the transactions of the directorates.

In a number of fields *local state bodies* have been established. As an important example may be mentioned the offices of the country's chief constables. Like the directorates, the local state bodies are usually headed by a civil servant assisted by a large or small hierarchically organised staff. As indicated by the designation, however, these bodies perform tasks only within a local area. Ministries and directorates—if any exist in the field in question—have the same powers *vis-á-vis* the local state bodies as have the ministries in relation to the directorates. In recent years the importance of local state bodies has dwindled because a growing number of tasks at local level are being assigned to the municipal councils. In the field of pollution control local state bodies now play a secondary part.

Besides the bodies mentioned so far, which form part of an overall hierarchy under the ministers' leadership, there are, in particular, two important categories of administrative authority with a more independent status. Those in one category are called by the common designation of *board*, although this covers several widely differing authorities. The boards have one thing in common; they are corporate bodies and the ministries (and directorates) have only such powers in relation to them as are sanctioned by law except that the boards are under an obligation to inform the ministry (and directorate) concerned about their activities. The members of the board may be experts within various provinces, representatives of interest groups, members of local authorities, members of the Folketing and judges or people who would qualify as such. There are local as well as central boards. Among the latter there are, in particular, a number of appeal boards, i.e. administrative tribunals dealing with complaints about the decisions of other administrative authorities with a view to possible amendment.

The other important group of independent authorities are the *municipal councils*, the local councils and the county councils. Since 1970 Denmark has been divided into 277 so-called primary municipalities in which a local authority numbering from 5 to 31 members is elected by the local population in general elections which are held every four years.

Prior to 1970 there were some 1400 municipalities which were divided into several categories, each with their separate rules of administration. Thus the number of municipalities has been considerably reduced and the principal rule is that a Danish municipality should at present have a population of at least 6–7000, which is considerably more than in most other West European countries. Besides the division into primary municipalities, the country is divided regionally into 14 counties in which there are elected, simultaneously with the municipal authority elections, county councils with from 13 to 31 members. Outside the division into counties, there are the municipalities of Frederiksberg and Copenhagen. Furthermore, a special arrangement applies to the metropolitan area—the municipalities of Frederiksberg and Copenhagen together with three surrounding counties—where a number of powers otherwise exercised by county councils and municipal authorities are vested in the Greater Copenhagen Council.[10]

The administration of municipalities and counties is controlled by municipal authorities and county councils, respectively. Under their control some of the functions are performed by standing committees, the mayor (who is elected by the municipal council for 4 years) and a considerable permanent staff. In the municipalities of Copenhagen, Aarhus, Odense and Aalborg several mayors and/or chief officers are elected who quite independently assume the functions that are elsewhere performed by standing committee.

Unlike the members of other municipal councils, the members of the Greater Copenhagen Council are elected indirectly by the municipal authorities in Copenhagen and at Frederiksberg, together with the three county councils that are covered by the arrangement.

The tasks of the municipal councils may be divided into two sets between which there is, however, no clearcut distinction. The first may be called the supply function. In order to fulfil the communal needs of the local population, the municipal councils may—and in many cases must—provide various services for their citizens. As examples there may be mentioned schools, social institutions, roads, public transport, recreational areas and refuse disposal. In the case of the county councils it will be a question of providing facilities like hospitals and colleges, which cannot usually be catered for within the individual primary municipality. The law provides a fixed framework for most aspects of the supply function of the municipalities, but the latter are, within certain limits, entitled to perform services for their citizens without statutory authority.

The supply function may be said to correspond with the traditional, fundamental idea which forms the basis of local autonomy. But besides their service functions, local councils to an increasing extent attend to the task of making decisions in respect of individuals, especially in the form of authorisations, orders and prohibitions. Among important sectors of this nature in which powers have been delegated to the local councils are building legislation, town planning legislation and the law of environmental protection. Whereas supply functions may be attended to without statutory authority, local councils must, like other authorities, be authorised by statute if they are to issue legal directives concerning the citizens.

In Danish law the municipalities are regarded as independent units whose freedom of action, like that of individuals, is protected by the rule of law. This means that the central authorities—or, as regards municipalities, county councils—may exercise the powers of a superior authority only to the extent authorised by statute. On the other hand, the Constitution requires that the municipalities shall be subject to a certain amount of state control although the nature of such control is left to be decided by the legislature.

State control of the municipalities varies considerably with the nature of the tasks and the legal framework for control is chiefly provided by the special legislation within the individual sectors. Most important in practice is the control exercised through general directives, plans, budgetary control, administrative appeal and guidance. Formerly central administration approval was required for a number of municipal transactions and this was an important means of exercising control; in some fields such arrangements are still important. Concurrently with the legislature's limitation of municipal transactions needing approval, ministries or directorates have been empowered in a number of fields to 'call in' a local authority ruling in order that it may be examined and decided at the central level. Besides the means of control mentioned which apply to certain departments, the Local Administration Act provides for general control of the municipalities. This forms a minimum basis for control and means, in particular, that some regional supervisory bodies and the Ministry of the Interior may intervene—if necessary with cancellation and the imposition of compulsory fines—in the event of illegal decisions being made by local authorities. The fact that general control of local authorities is at a minimum level is apparent from the fact that if there is a more extensive control within a field, e.g. an arrangement for administrative appeal under special legislation, then the general control cannot as a rule be exercised.[11]

1.2 SOURCES OF LAW IN THE FIELD OF ENVIRONMENTAL PROTECTION

Questions as to which sorts of material to consider when making statements concerning the legal position and how to use such material are basically common to all dogmatic legal disciplines and are dealt with under the headings of 'jurisprudence' or 'legal philosophy'.[12] Special problems related to sources of law arise in connection with individual subjects, e.g. constitutional law, administrative law and criminal law, and these problems are dealt with in the general textbooks within the respective disciplines.[13] In this study there is no reason to deal with questions of principle as regards sources of law. It is only intended to provide a practical survey of the types of material which are most significant for the ensuing sections of the study.

In principle, the most significant basis for statements concerning the legal position relating to pollution control in Denmark is to be found in *acts* of parliament, i.e. regulations adopted by the Folketing, approved by the Government and promulgated in the official publication *Lovtidende*. Many different acts are important, but the central act concerned with the subject is Act No. 372 of 13 June 1973, the Environmental Protection Act.[14]

The acts often grant the minister powers to regulate various conditions within certain limits, particularly as regards pollution control. This means that significant contributions towards determining the state of law must be looked for in *general directives laid down administratively*. Such directives are issued either by a minister or by the Environment Agency. If the directives are to be immediately binding on the citizens they must, like acts, be published in the *Lovtidende*. In legal theory these directives are referred to as *anordninger* (ordinances) but they usually appear in the *Lovtidende* under the heading *bekendtgørelse* (notice). Where the rules cover a very wide field the designation *reglement* (regulations) is sometimes used, however, cf. Government Notice No. 170 of 29 March 1974 concerning regulations on environmental protection. Where the sole purpose is to bind subordinate authorities as regards their administration of the Act it is not necessary to publish the regulations in the *Lovtidende*: it is sufficient to inform the subordinate authorities and the regulations will usually be printed in another official publication, the *Ministerialtidende*, as *cirkulærer* (circulars). These general directives may also be issued as notices, which is often done.

Prior to the Environmental Protection Act which took effect on 1 October 1974, regulations concerning only one municipality and laid down in collaboration with the relevant local authority played an important part. Although considerable standardisation was achieved on the initiative of the Ministry, the legal position varied in many respects from municipality to municipality. This situation was changed radically by the Environmental Protection Act which only provides for the issue of local rules within the province of the Environmental Protection Regulations, i.e. concerning matters which are generally of less importance to environmental protection.[15] Such local directives will usually appear under headings like *regulativ* or *vedtægt* (by-laws) and they are published in the individual municipalities, often in the local press.

Also the decisions on concrete cases made by public authorities are important as a source of law because their existence will greatly influence future decisions on similar issues.

In principle, *case law* in the form of previous court decisions on concrete cases is the most important source of law. Nearly all Supreme Court judgments, a wide selection of judgments made by the two High Courts of Justice and a few passed by inferior courts are published in the *Ugeskrift for Retsvæsen* (legal weekly) (hereafter abbreviated to *U*). In the province of environmental protection there are few decisions of any interest, however. This is probably due chiefly to the fact that the authorities' powers to decide concrete cases are by and large discretionary. The passing of the Environmental Protection Act adds another factor in that this Act provides for a special appeal board with a certain judicial stamp which means, on the one hand, that an unbiased hearing by a superior authority is possible and, on the other, that the chances of having the final administrative decision overruled by the courts will usually be small.

This means that from a practical point of view interest centres on *administrative practice*. Fairly comprehensive summaries of a considerable number of decisions made by the Environmental Appeal Board are published in the *Kendelser om fast ejendom* ('Decisions on real property', a publication abbreviated below to *KFE*) which appears in four volumes each year.[16] The transactions of the Environment Agency are not usually published. A few booklets in its current series *Nyt fra Miljøstyrelsen* ('News from the Environment Agency') on the administration of the Environmental Protection Act contain a summary of some decisions which are considered fundamental by the Agency.

All the sources of law mentioned above are by way of binding decisions, general or concrete. Most interesting in a study of the legal regulation of environmental protection are a number of *guidelines* issued chiefly by the Environment Agency. Some of them contain a general report on regulations in acts and *notices* and in some respects contribute to their interpretation. Other guidelines contain recommendations or suggestions for the application of discretionary provisions in acts or *notices*. Often the guidelines combine these two elements. Some instructions—notably those which interpret acts or notices—are printed in the *Ministerialtidende*, usually under the heading of *cirkulære*. In other words, a 'circular' may be partly an instruction, partly a memorandum containing various recommendations. Other guidelines are issued separately by the

Environment Agency in a current series *Vejledning fra Miljøstyrelsen* ('Guidelines from the Environment Agency').[17]

Legal literature on issues concerning environmental protection is as yet rather scarce compared with the literature on various other sectors of law but it is growing rapidly after the environmental reform of 1973/74. Among major publications may be mentioned Niels Borre's annotated edition of the Environmental Protection Act, 1973, the first edition of the present study from 1976 (the English version), Claus Haagen Jensen's *Godkendelse af forurenende anlæg*, 1977 (stencilled) and Claus Haagen Jensen's and Claus Tønnesen's 'Miljøbeskyttelse' in *Dansk Miljøret*, vol. 3, 1977. Furthermore, there are papers covering other subjects as well as environmental protection, e.g. Bent Christensen in *Dansk Miljøret*, vol. 1, 1978, and also a series of articles which will be mentioned in connection with the individual sections below.

The state of pollution in Denmark was described during the years 1970–72 in 29 publications issued by the *Forureningsrådet* (the Pollution Council) which has now been abolished. Studies of this nature which deal only peripherally with the legal rules concerning environmental protection have been subsequently followed up by the Environment Agency which has published the results in a number of publications. A concise survey of the effects of environmental reform was published in November 1979 by the Environment Agency in the book *Miljøreformen* ('The Environmental Reform').

1.3 CENTRAL AUTHORITIES

At the central level three administrative authorities in particular deal with matters relating to environmental protection, namely the Ministry of the Environment, the Environment Agency and the Environment Appeal Board. With the Ministry of the Environment as the superior authority, the first two deal with a broad spectrum of tasks, whereas the Appeal Board's competence, within which it is the supreme administrative authority, is rather limited. In addition to the three bodies mentioned there is the Environmental Credit Council which administers the Environmental Investment Subsidies Act and some other specialised authorities which will be dealt with in Chapters 2–10.[18]

The Ministry of the Environment is an authority of the 'ministry' type described in 1.1.2 above. Only one of the Ministry's six offices is concerned exclusively with pollution control whereas the others deal with matters such as physical planning and nature conservation. The staff of the Ministry dealing with environmental protection matters is rather small, totalling some 20 persons.

Through legislation wide powers are vested in the Ministry of the Environment. First and foremost the Ministry is empowered to issue general rules on various matters. This is done under the provisions of the Environmental Protection Act[19] and under the provisions of the special acts relating to environmental protection, e.g. the Limitation of the Sulphur Content etc. of Fuel Act and the Chemicals and Chemical Products Act. The Ministry has thus been granted far-reaching powers to regulate the conduct of both citizens and authorities within the field of environmental protection. From the point of view of constitutional law such extensive delegation of legislative power to a minister is not unobjectionable. It is interesting that when commenting on the Bill the Minister said about section 6 of the Environmental Protection Act (which deals with the power to lay down emission limits, etc.) that proposals for particularly radical measures would be submitted to the Folketing in each case in the form of a bill notwithstanding the fact that there might be authority to implement the regulations by issuing a notice. As an example illustrating that this declaration by the Minister is observed in practice, it may be mentioned that binding regulations concerning the lead content of petrol were laid down by a special act although the wording of the Environmental Protection Act provided the necessary authority for issuing such regulations.[20] It should be added in this connection that during the reading of the Environmental Protection Bill in the Folketing the Minister promised to keep the parliamentary Environmental Committee informed on the more important steps concerning emission limits, etc.[21]

Furthermore, in several fields the Ministry has the original authority under the Act to make concrete decisions[22] and is also entitled to intervene in subordinate authorities' handling of actual cases or in connection with their decisions in individual cases.[23] With the present modest staff at its disposal, the Ministry would be quite unable to attend to all these tasks. The legislation therefore enables the Ministry to delegate some of its powers to other authorities or at least to obtain assistance from them.[24] This means that the more detailed definition of the Ministry's work is undertaken by the Ministry itself.[24] This legislative arrangement has been chosen in order, on the one hand, to ensure the Minister's influence on the administration of the current rules and on the planning of new initiatives against pollution and, on the other hand, to allow the Ministry to delegate all powers not necessary for purposes of control to other authorities.

The Environment Agency is a directorate under the Ministry of the Environment (cf. 1.1.2 above on directorates). The Agency deals exclusively with environmental protection and has a staff numbering some 150 headed by a director. Among the members of staff there are doctors (hygienists), engineers, biologists, chartered surveyors, lawyers, political scientists and economists.

The functions of the Agency may be considered to fall into five categories. First, the Agency advises the Ministry on its duties, in which connection it should be noted that the Agency generally possesses the greatest technical expertise. This applies not only to the drafting of general regulations, but also in relation to individual cases or projects, see section 45, subsections 1 and 2, of the Environmental Protection Act. Second, the Agency acts in an advisory capacity to authorities other than the Ministry, viz. local authorities, see section 45, subsection 1, of the Environmental Protection Act. Third the Agency exercises numerous powers which under the Act are vested in the Ministry but which have been delegated by the latter to the Agency. On the subject of delegation of powers to the Environment Agency, Government Notice No. 178 of 29 March 1974 should be referred to, in which it is laid down that the power to issue recommended standards for emissions, etc. has been transferred to the Agency. Fourth, the Agency is generally the body of appeal regarding decisions taken by local authorities.[26] Finally, most of the public information service is handled by the Agency.

Within its field of activity the Environment Agency is entirely subordinate to the Ministry. The latter may request information from the Agency and can, on its own initiative or after hearing complaints, reverse decisions taken by the Agency, take over undecided cases from the Agency and issue directives to the latter. Where the Environment Agency has reached a decision which may, in pursuance of section 76 of the Environmental Protection Act, be appealed against to the Environmental Appeal Board, the case cannot, however, be submitted to the Ministry of the Environment for re-examination[27] and the Minister may not alter the decision on his own initiative.[28] On the other hand, under section 70, subsection 3, of the Environmental Protection Act, the Ministry is entitled to take over cases which have not yet been decided by the Agency.

On the subject of relations between the Ministry of the Environment and the Environment Agency it should be further mentioned that the cooperation between them is closer than is usual between a ministry and its directorates, with the result that the powers mentioned will rarely be formally applied. This cooperation manifests itself among other things in frequent meetings between the permanent heads of the Ministry and the Agency, in the frequent exchange of staff between the two authorities and in the setting up of project groups for the study of specific major problems in which staff from both bodies participate. Although a matter should officially be dealt with by one of the two authorities, it will thus very often be considered by both, either because the Ministry asks the Agency for assistance or because the Agency submits questions of principle to the Ministry. In practice, the two authorities operate to a great extent as one unit.

The Environmental Appeal Board was established under head 12 of the Environmental Protection Act. The Board comprises, on the one hand, a president who must fulfil the general conditions for being a Court of Appeal judge and one or more deputies who must fulfil the general conditions for being a judge and, on the other, some appointed expert members, their number being decided by the Minister. The expert members are appointed by the Minister for a period of 4 years at a time on the recommendation of some trade organisations, viz. the Federation of Danish Industries, the Joint Council of Danish Farmers' Cooperative Societies, the Federation of Danish Agricultural Societies and the Federation of Danish Smallholders' Societies jointly and of the Environment Agency. It is assumed that the trade organisations will preferably propose members with expert knowledge of business economics and production matters, whereas the members proposed by the Environment Agency will be experts on matters of environmental technology, ecology and the like. The latter are appointed for each individual case by the President, who appoints an equal number of members from the groups of persons proposed by the trade organisations and the Environment Agency, respectively. This procedure enables the President to select appointed members with special qualifications for assessing the circumstances of the individual case in question.

As the name suggests, the function of the Environmental Appeal Board is limited to dealing with complaints arising from concrete decisions. Not all decisions on environmental protection may be referred to the Board, however. Only where expressly provided by statute or regulation may decisions be referred to the Board and this possibility of re-examination is generally reserved for the most important decisions on environmental protection which have been dealt with in advance by the Environment Agency or, in exceptional cases, by the Ministry instead of the Agency.

The main provision is section 76 of the Environmental Protection Act which lays down that the decisions which may be referred to the Board are mainly those concerning specially polluting undertakings (specified in an appendix to the Act), orders or prohibitions for the prevention of pollution of groundwater, permission for sewage discharge into watercourses, lakes or the sea and orders regarding existing sewage installations where such orders affect a municipal or industrial sewage system. It should be noted that within the sphere of the Environmental Protection

Act the Environmental Appeal Board has no jurisdiction to review important municipal decisions on plans concerning sewage. This is explained in the preparatory works of the Act by a reference to these issues as being of a 'general' nature. The Board was considered qualified to deal with decisions on legal and technical issues, whereas it was considered unqualified to deal with cases in which the decision involved a political responsibility.[29]

The chief functions of the Board fall within the sphere of the Environmental Protection Act but as regards other legislation the Board has also been given jurisdiction as a superior appeal instance. This jurisdiction applies to the Sulphur Content etc. in Fuel Act, the Disposal of Oil and Chemical Waste Act, the Water Supply Act, the Chemicals and Chemical Products Act, the Recycling Act, the Maritime Environment Act, the Raw Materials Act and the Zoning Act. Further it should be noted that with a different composition as regards the appointed members of the Board, powers were conferred on the Environmental Appeal Board by the Raw Materials Act of 1977.[30]

Pursuant to section 77, subsection 3, of the Environmental Protection Act, the Minister of the Environment can lay down further rules regarding the work of the Environmental Appeal Board, including rules concerning its procedure. Under this provision, the Minister has issued a number of formal directives, concerning mainly the rules of procedure of the Board.[31] The Minister can, moreover, influence the composition of the Board to some extent in that he or she decides which persons are to be appointed members. In this connection, it should be recognised that, as regards members appointed, the Minister is under no obligation to comply with the recommendations of the Environment Agency and the trade organisations. Otherwise the Board itself is independent and so are its individual members and it cannot, in particular, accept instructions regarding handling and decision in individual cases.[32]

The composition of the Environmental Appeal Board may give, and has already given, rise to some critical remarks.[33] The composition of the Board can be said to be biased because some members are appointed who must be assumed to represent the polluters' interests (the members recommended by the trade organisations) whereas there are no members representing those who are exposed to pollution. In response to this, it may be said that the members recommended by the trade organisations must be experts, not representatives of interest groups as is shown by the fact that the Minister is empowered to appoint all members in this category. The weight to be attached to this consideration seems limited, however. It is hard for members recommended by trade organisations not to identify themselves to some extent with the interests of the industrialists in limiting the demands based on environmental protection considerations. It is difficult to assess whether this imbalance in the composition of the Board has manifested itself in practice and, if so, how. Yet there does not seem to be any definite trend towards modification by the Board of environmental protection requirements laid down in the Environment Agency's decisions. It has happened in some cases, but then the Board has also made the environmental protection requirements more stringent in a considerable number of other cases and even revoked approvals granted by the Environment Agency.[34]

The Environmental Credit Council was set up in pursuance of the Environmental Investment Subsidies Act, Act No. 682 of 23 December 1975. It consists of a chairman, 2 representatives of the Ministry of the Environment and 6 other members who are all appointed by the Minister of the Environment. The '6 other members' are appointed on the recommendation of the Ministry of Commerce, the financial institution organisations, the agricultural organisations, the Federation of Danish Industries, the Economic Council of the Labour Movement and the municipal organisations. The Environment Agency acts as secretariat to the Environmental Credit Council.

The Council's brief is to decide cases concerning government-subsidised private investments which will result in considerably reduced pollution by the individual undertakings. Originally this support was rendered primarily by way of interest subsidies so as to reduce the rate of interest on loans, but following an amendment to the Act in 1978 the support is now granted exclusively as direct subsidies. Promises of subsidies may not be made after 31 December 1980.[34a] The decisions of the Council cannot be referred to any other administrative authority. The Minister of the Environment can lay down, and has laid down, specific rules for the Environmental Credit Council's operations including rules governing its procedure.[34b]

1.4 REGIONAL AND LOCAL AUTHORITIES

At the regional and local levels there are primarily two kinds of body with the power to make decisions in the field of environmental protection—the county councils and the local councils; generally the duties are shared between them so that decisions are made either by a county council or by a local council. Furthermore, binding decisions of limited scope are taken by Agricultural Land Tribunals and it is incumbent on certain local bodies to assist the county councils and the local councils in the exercise of their functions.

As many duties as possible which are not performed centrally are delegated to the *local councils*. In regard to the number and composition of the latter, reference should be made to 1.1.2 above. The geographical division of the country into various municipalities is apparent from Notice No. 62 of 27 February 1970 with many subsequent minor amendments.

It is obvious that the local council as such cannot itself attend to all the functions delegated to it by legislation. The local council can, in principle, only supervise the current administration which is carried out on its behalf, in particular, by standing committees, mayors and the permanent staff. This also applies to matters of environmental protection. As regards this sector, it should be mentioned that it is possible to set up standing committees concerned solely with local environmental protection issues. Most local councils, however, have delegated matters concerning pollution control to standing committees which also attend to other tasks, e.g. town planning, public works and sewerage. As regards committees, the local councils have otherwise organised themselves in various ways. It should be specially mentioned that some local councils have shared the environmental protection tasks between two different standing committees, which generally does not seem expedient.[35]

The basic principle for allocating environmental protection tasks at the regional or local level is that as many functions as possible should be delegated to local councils. To the citizens this means that the local council is generally their first resort in cases concerning pollution control. Among the local council's most important tasks in the field of environmental protection mention should be made of the following:

(1) Power to sanction enterprises with particular pollution problems unless, as an exceptional measure, these tasks have been delegated to the council council.[36] Orders and prohibitions in connection with such enterprises are normally issued by the local council, even if the undertaking in question is sanctioned by the county council, cf. section 44 together with section 42 of the Environmental Protection Act.

(2) The power to authorise cesspools and tanks for domestic sewage.[37]

(3) Permission to lead sewage to public sewage systems, to the systems of some small private domestic sewage installations and, in exceptional cases, to discharge industrial waste, etc. into watercourses, lakes or the sea.[38]

(4) Administration of the provisions of the general environmental protection regulations (concerning, for example, refuse, dumps, beaches, swimming pools and enterprises or activities which cause a moderate nuisance).[39]

(5) Supervision of the observance of, *inter alia*, the Environmental Protection Act and regulations issued in pursuance of the latter, cf. sections 48–49 of the Environmental Protection Act and Notice No. 177 of 29 March 1974.

(6) Setting up most of the major sewage systems, including plants for the purification of sewage, cf. section 21 of the Environmental Protection Act.

The *county councils* too carry out important functions within the field of pollution control. Regarding the number and composition of the county councils, reference should be made to 1.1 above. The powers of the county councils to delegate their duties to committees, mayors and permanent staff are similar to those which are vested in the local councils. The areas of jurisdiction of the county councils are apparent from Notice No. 62 of 27 February 1970 mentioned above, with subsequent minor amendments.

The tasks of the county councils in the field of environmental protection can be split up into two groups. In the first place the county councils perform functions which cannot reasonably be carried out by the local councils. This may be due to the fact that the scope of the matter extends beyond the local boundaries, e.g. the discharge of sewage into watercourses or the sanctioning of enterprises which, like cement works, are capable of causing considerable air pollution.[40] The reason may also be that the polluting plant in question is operated by a local council.[41] In addition, the county councils have an important role to play in the planning of environmental protection. Thus the municipalities' plans for enlarging sewage systems are subject to approval by the county councils, cf. section 21 of the Environmental Protection Act, the county councils have to prepare surveys of sources of pollution in the county containing, among other things, proposals on the future siting of enterprises for which, owing to the risk of pollution, special requirements as to sites have to be made, so-called 'recipient quality planning'.[42] The county council is also responsible for plans for the enlargement of control units with laboratories.[43]

On the other hand, county councils do not normally perform administrative functions in relation to the local councils. Thus the county councils have no authority to give instructions to local councils and they cannot reverse the decisions taken by the latter. Apart from the county councils' duties as regards planning, administration is considered the responsibility of the central authorities, first and foremost the Ministry of the Environment and the Environment Agency. In other words, a general characteristic of the distribution of tasks between local councils and county councils is that the two types of body have their separate

REGIONAL AND LOCAL AUTHORITIES

functions in the first instance. The legislators have endeavoured, in this as in other fields, to avoid 'double administration'.

As far as the metropolitan area is concerned, i.e. the municipalities of Copenhagen and Frederiksberg and the counties of Copenhagen, Frederiksborg and Roskilde, a special arrangement applies. Powers which are normally vested in county councils are as a rule exercised in this area by the *Greater Copenhagen Council* which is indirectly elected by the members of the local councils in Copenhagen and Frederiksberg and the county councils of Copenhagen, Frederiksborg and Roskilde.[44]

The relationships between central authorities and the three different types of council within the field can be dealt with collectively. All three types of bodies are municipal councils, which means that other authorities have only such power over them as is sanctioned by law. For the Ministry of the Environment and the Environment Agency certain opportunities of exerting influence on local councils' administration in the field of pollution control are provided by the Environmental Protection Act. The most important thing is that the Ministry (or, by delegation, the Agency) can lay down general directives as regards the activities of local councils and county councils. These may relate partly to the content and premises of the decisions,[45] partly to the procedure of local authorities[46] and also to supervisory and control operations.[47] Secondly, appeals against decisions taken by municipal councils under the Act may generally be made to the Environment Agency which is empowered to change the decisions in question.[48] Thirdly, the Minister and, in the event of delegation, the Environment Agency are empowered on their own initiative to intervene in individual cases. This may be done by the Minister taking over a case from a local council and deciding it himself, section 47; by the Minister ordering a local council to take up an issue for assessment and decision, section 57; and by the Minister, *ex officio*, changing a decision taken by a local council, section 70, subsection 3. It should be emphasised that both the power to 'call in' a case and the power to change a decision on his own initiative are subject to stringent conditions. In addition to the influence exercised through these formal superior powers, the Ministry of the Environment and the Environment Agency can exert considerable influence on the activities of local bodies by way of guidance given generally or in concrete cases.

A considerable number of general directives have been issued which are binding upon the local councils in their administration of the environmental protection legislation, but the majority of these directives have related to the competence and procedure of local bodies. The Minister's power to intervene in actual environmental cases on his own initiative has not yet been exercised. In practice, control has been chiefly exercised through the issue of general guidelines and through the system of appeal.

The general guidelines have a great effect in the municipalities both by reason of the central authorities' expertise and in the light of their fairly extensive formal administrative powers. In the nature of things a local council tends to follow general directions issued by the Environment Agency if it knows that its decision in a specific case may be referred to that same Environment Agency.[49]

Among the decision-making bodies, mention must also be made of the *Agricultural Land Tribunals*. Agricultural Land Tribunal is the common name applied to three types of administrative board with a judicial procedure, viz. Agricultural Board, Agricultural Commissions and Higher Agricultural Commissions. The Agricultural Board normally operates within the area of the district court in question. The province of the Agricultural Commission is, in principle, the county, but a number of counties are divided into several districts, each with its own Agricultural Commission. There are six Higher Agricultural Commissions operating in areas determined by the Minister of the Environment. An Agricultural Land Tribunal has a permanent president who is, in the case of the Agricultural Commission, one of the district court judges of the county in question and, in the case of the Higher Agricultural Commission, a person who fulfils the normal requirements for appointment as an Appeal Court judge. Besides the president, the Agricultural Land Tribunal consists of two, and the Higher Agricultural Commission of four, members who are appointed for each individual case and who are usually members of a local council. Agricultural Land Tribunals may and usually do employ technical experts. Cases usually start with an Agricultural Board (particularly minor cases) or Agricultural Commission and their decisions may be appealed against to an Agricultural Commission or a Higher Agricultural Commission, respectively.[50]

Formerly, the Agricultural Land Tribunals had important duties within the field of environmental protection, particularly in connection with drainage installations. By the amendment of acts in connection with the passing of the Environmental Protection Act, the functions of the Agricultural Land Tribunals in the field of pollution control were restricted to decisions on questions of compensation for intervention for the purpose of protecting the environment.[51] Cases concerning the pollution of watercourses, lakes or the sea, including cases relating to the approval of drainage installation projects, which were, at the time of coming into force of the Act, being dealt with by an Agricultural Land Tribunal might, however, until 1 December 1976 be decided by the latter. If the case had not been completed by then, jurisdiction would pass, as a rule, to the authority of first instance which was competent under the Environmental Protection Act—usually a municipal council.[52]

Many cases relating to environmental protection contain complex issues particularly of a technical, hygienic, biological, economic and legal nature. The ability of the local authorities to perform their duties satisfactorily therefore depends very largely on their being able to secure the necessary expert assistance. Some can be found among the municipality's own permanent staff. Thus most municipalities employ one or more engineers and the larger municipalities generally also employ people with economic or legal training as well. However, the municipal authorities have to draw to a considerable extent on *specialist help* from outside their own administration.

The *local food control units* are important in this context. Part 9 of the Foods Act No. 310 of 6 June 1973 provides for the establishment of one or more food control units comprising laboratories and supervisory staff. The detailed arrangements for each county are set out in a plan prepared by the county council and approved by the Minister of the Environment. This includes decisions made relating to the geographical siting of the unit, the area to be served by it, the authority which is to operate it and the extent to which the unit is to execute laboratory work on environmental matters.[53] The units are normally operated by a single municipal council or by several jointly, but it is possible to set up units to be operated by the county councils. Within the field of the Environmental Protection Act these units are usually only able to undertake examinations of water, particularly bacteriological examinations.

In matters of hygiene the municipal councils are able to obtain expert assistance, especially from the *government medical officer agencies.* In each county there is a medical officer agency with one or more medical officers and their assistants. The duties of the medical officer agencies are, among other things, to give advice to government, county and municipal authorities on medical, hygiene, environmental and welfare matters.[54] In this connection, it may also be mentioned that medical officers have an important independent position within the field by virtue of being entitled under section 74 of the Environmental Protection Act to appeal to the Environment Agency against decisions made by municipal councils, cf. also 1.6.2.3 below.

Hitherto *the police* have contributed significant information in a number of cases concerning environmental protection, e.g. through interrogations and inspections. This situation will change as a result of the Environmental Protection Act as in future the police will only be called in on matters concerning environmental protection with a view to investigating criminal cases.[55]

These facilities for obtaining assistance are not always sufficient. They will not, for example, suffice where a municipal council needs to have measurements of noise or air pollution analyses carried out. In these and similar cases the councils will have to rely on assistance rendered by central government authorities, e.g. the Environment Agency or the State Food Institute, by scientific institutions or consultant engineering firms and the like.

In this connection it should be mentioned that a considerable part of the work to clarify the nature and extent of industrial pollution and the appropriate preventive measures must be carried out by the enterprises causing the pollution. This clearly applies in the case of enterprises which are subject to prior approval. If the application of the firm in question gives reason to anticipate pollution without this being clarified by the applicant, approval may be refused until such time as the undertaking has itself proved, if necessary by means of technical investigations, that the pollution will not exceed acceptable limits. In cases involving orders/ injunctions, cf. especially section 44 of the Environmental Protection Act, the authorities must prove that there is unacceptable pollution but need not in an order specify the preventive measures to be adopted. The firm in question must itself discover how the nuisance can be reduced to an acceptable level and have the investigations necessary for this purpose undertaken.

1.5 INDEPENDENT CONSULTATIVE BODIES

Independent consultative bodies with obligations or rights under the Environmental Protection Act do not exist. In one particular field the Advisory Board dealing with measures against oil pollution which was set up in pursuance of section 9 of the 1956 Act concerning measures against oil pollution may be mentioned here.[56] It consists of a chairman, appointed by the Minister of the Environment, and representatives of interested authorities and organisations as decided by the Minister. Further information about the composition and functions of the Advisory Board is given in 1.6.1 below and in Chapter 5. See too 1.4 above on the opportunities for obtaining expert help in matters of pollution control.[56a]

1.6 RIGHTS OF SPECIAL INTEREST GROUPS

In almost all fields subject to public control there are various associations working in the interests of individuals who are affected. Within the environmental protection sector too there are various interest groups, groups which look after the interests of polluters as well as associations of people who are or may be exposed to pollution.

The term 'interest groups' covers two kinds of association between which it may be difficult to distinguish in practice. Under the heading of interest group fall first and foremost the traditional 'interest organisations' such as the trade organisations and the trade unions. It is characteristic of these, in particular, that their aims are comparatively general, their structure is very rigid and formalised, and the support of their members is predominantly motivated by expectations of material advantages. Secondly, interest groups comprise the so-called 'grass-roots or basic movements' of which a considerable number have been formed during the past decade, especially at the local level. The objective of such movements is typically very narrow—it may be to prevent or ensure the implementation of a particular project; the organisational structure is very loose and informal, and the prospect of material advantages is normally of little importance as a motive for supporting the interest group.[57]

Interest groups may have certain rights within a particular field of administration. Some of the most important rights of interest groups in the field of environmental protection will be illustrated below.

1.6.1 Representation on administrative bodies

As is apparent from 1.3 and 1.4, the administration of the Environmental Protection Act is mainly in the hands of authorities where interest groups are not represented. However, it may be said that industrial as well as agricultural organisations enjoy representation of a kind on the Environmental Appeal Board (see 1.3). There is a definite representation of interests, however, on the board mentioned above under 1.5 concerned with measures against oil pollution, to which are appointed, besides the chairman, representatives of the Danish Shipowners' Association, the Association of Ship and Engine Builders in Denmark, the Shipmasters' Association in Denmark, the Engineers' Association, the Joint Committee of the Oil Trade and of a number of public authorities. The brief of this body is to follow the development of the problem of oil pollution, recommend practical measures for the prevention of oil pollution, and otherwise assist the Minister in enforcing the Act concerning measures against oil pollution of the sea. Furthermore, the Board can make recommendations to the Minister in accordance with section 11 of Act No. 290 of 7 June 1972 concerning measures against pollution of the sea with anything other than oil, cf. Chapter 5 for details.

1.6.2 The right to be consulted

Under Danish law, organisations are only entitled to have a case put before them and to give their opinion on it prior to an administrative decision, general or concrete, being taken, if such a right is provided for in the law or in regulations issued in accordance with it. Many acts contain provisions of this nature. By way of example, section 9 of the Chemicals Act may be mentioned, according to which the Minister of the Environment has to negotiate with 'the national trade and consumer organisations most concerned', including the trade unions and the employers association, and the municipal organisations prior to the issuing of rules under the statute. Such provisions are to be found only to a very limited extent within the field of environmental protection as it is delimited in this study. It should further be noted that in accordance with section 20 of the Environmental Protection Act, section 8 of the Sewage Notice[58] provides for the convening of a meeting of interested local organisations at which the sewage plan of the local council is discussed. Furthermore, it should be mentioned that, according to subsection 2 of section 18 of the Environmental Protection Act, the municipal organisations have the right to express opinions on directives which delegate to the municipal councils the county councils' powers to allow sewage discharge.

Even if the interested organisations are not legally entitled to be consulted, some of them will, in fact, get an opportunity to express their views on projected general directives before such directives are implemented. Thus the preparatory works of section 6, in particular, of the Environmental Protection Act, which empowers the Minister of the Environment to issue general directives especially concerning discharge by polluting undertakings, stipulate that the trade organisations shall be involved in the working out of such regulations,[59] and, in practice, under the present as well as previous legislation interested trade organisations have been given an opportunity to express their views. Municipal organisations too—the National Association of Municipalities representing the local councils and the Association of County Councils in Denmark representing the counties—are usually given an opportunity to express their views on general directives under consideration.

The trade organisations will, on the whole, represent the interests of polluters even if a positive interest in comprehensive measures for environmental protection may predominate in some trade organisations, for example, the Associations of Deep-sea Fishermen in Denmark. While business interests are tightly and strongly organised, both in specialised trade associations and larger associations, the organisation of direct interest in the protection of the environment in one or more respects displays little strength or breadth. In the field of environmental protection there are no general 'consumer organisations' such as are known, for example, in relation to food and monopoly legislation. Only in a few, rather specialised, fields are there at present nationwide organisations representing what might be called 'environmental interests'. The most notable one is probably the organisation known as NOAH, which is an association influencing public opinion and working to disseminate a knowledge and understanding of ecological problems. This association has a very limited membership and its activities are chiefly carried on by the members of groups of staff and students at the institutions of higher education. Other organisations watching over general environmental interests are the Danish Society for the Preservation of Natural Amenities whose principal activities fall within the scope of the Nature Conservation Act and the Outdoor Activities Council which is a joint representation of outdoor-life organisations in Denmark. The Anglers' Association in Denmark may be mentioned as an example of a nationwide association representing narrower environmental interests, which has made its mark increasingly in recent years. The rather limited and comparatively weak organisation of environmental interests has meant that the central authorities' consultation procedure prior to issuing general directives usually only involves trade organisations and municipal associations.

1.6.3 Other rights

Among these mention will be made first of the opportunities open to organisations to obtain information about the material which forms the basis of administrative decisions. In this respect there are no particular rules in the environmental protection sector but, as a rule, under the Act on Public Access to Documents of the Administrative Authorities,[60] organisations can, like any member of the public, request permission to see all documents in cases dealt with by administrative authorities if they are able to 'indicate' the case to the authority in question, that is to say they must have some prior knowledge of its existence.[61] The right to see documents includes matters concerned with general directives and also with concrete decisions, but a lot of documents and information will be exempted, of course. This applies, for instance, to the authorities' internal material, information on the personal and financial affairs of individuals and on production and business matters, if it is important to the person or firm in question that the information should not be made public. The general principle of access to documents only implies a right to acquaint oneself with documents, if need be by paying for copies, but the organisation cannot ask for a decision to be postponed until it has had an opportunity of expressing an opinion, see 1.7.2 below on access for 'interested parties' to inspect documents.[62]

The provision concerning secrecy contained in section 85 of the Environmental Protection Act has been a particular problem in regard to access to documents. According to the Public Access to Documents Act, special provisions on secrecy contained in the legislation, that is to say provisions concerning secrecy outside the Danish Criminal Code and The Danish Civil Servants Act, entail that information covered by such special provisions may not be given to parties other than those involved in the case in question. Section 85 of the Environmental Protection Act on secrecy was originally—although probably unintentionally—such a special provision on secrecy, which led to considerable restriction of the public's opportunities of obtaining information concerning documents on environmental matters.[63] However, by Act No. 107 of 29 March 1978, section 85 was amended so that the provisions on secrecy will no longer in itself restrict access to documents on environmental matters.[64]

Secondly, organisations representing different kinds of interests in connection with environmental protection legislation are entitled, in certain circumstances, to refer an administrative decision to a superior administrative authority or a court. However, in Danish law these questions are best dealt with in conjunction with those concerning the individual's rights of appeal and action, see 1.7.3 and Chapter 5. The same applies to the rules governing the claims of associations to be informed of decisions made, a subject closely related to the right of appeal.

1.7 RIGHTS OF THE INDIVIDUAL

The opportunities for citizens to influence the administration of legislation related to environmental protection fall into three categories. In the first place the individuals affected have certain powers before a decision is made. This matter is dealt with in 1.7.1 and Chapter 2. Secondly, individuals are entitled to appeal to superior administrative bodies against decisions made by inferior authorities, see 1.7.3 and Chapter 4. Finally, individuals may bring actions in the courts concerning matters of environmental protection in order to have an administrative decision set aside, to obtain an injunction ordering another private individual to abate a nuisance or to obtain damages, cf. 1.7.5–7 below.

In general, as regards the individual's rights, it is observed that they concern concrete cases whereas organisations, as already mentioned, can exert influence in matters concerned with the issue of general directives. What is written in 1.7.5 below on the right of action, however, will also touch on the possibility, during a lawsuit, of contesting general rules laid down administratively.

1.7.1 The right to initiate administrative proceedings

Where controls are based on a system of permits, it goes without saying that proceedings are initiated by individuals through the filing of applications. It is doubtful, on the other hand, if by application to an administrative authority a citizen can initiate proceedings for an order or prohibition against another individual, thus legally obliging the authority to examine and decide the case. Rules on orders and prohibitions are enforced *ex officio* by the authority in question. This means that, generally speaking, it will examine the case and take a decision—positive or negative—on the basis of an application. With the generally informal administrative procedure in Denmark, it is only very rarely that the question of whether the citizen concerned is entitled to have a particular matter dealt with on its merits is taken to extremes. As a general rule— in the field of environmental protection too—the position is that any person who has an individual and essential interest in a decision has a right to initiate administrative proceedings on the issue in the sense described above. Concerning the notion 'individual and essential interest' readers are referred to 1.7.3 on the right to appeal.

1.7.2 The right to be heard, etc.

In legal theory this issue is usually referred to as the right of the individual to 'contradiction'. This means a right open to any individual who will be affected by a decision in a qualified way to acquaint himself with the facts on which a decision is to be taken and to express his opinion prior to a decision being taken by the authority.

Such a right of 'contradiction' presupposes three things. In the first place, the citizen must be aware or learn of the existence of a case which affects him and perhaps, in addition, facts have come to light making it important for him to take a stand. Secondly, he must have access to factual information available to the authorities which means, in particular, to documents in the case. Thirdly, he must be entitled to apply for the decision to be postponed until he has had an opportunity to make a statement. These last two conditions of the right to 'contradict' are mainly governed by the general provisions in part 2 of the Public Access to Documents Act on access for 'interested parties' to which nothing of importance has been added by the legislation on environmental protection. There are no general statutory provisions covering the first of the three conditions, but some parts of the problem are governed by provisions in the Environmental Protection Act and general directives issued in pursuance thereof.

According to part 2 of the Public Access to Documents Act, the 'interested parties' in concrete cases have certain powers which extend beyond the public's right under the provisions of part 1 of the Act to see public authority documents. For one thing, the parties have a right to see more documents than the public in general as the exceptions under part 1 are substantially limited in relation to interested parties (see 1.6.3 above) and they are also entitled to free copies. Furthermore, the parties involved can in the main request that a decision be postponed until they have made a statement in the case. This rule also applies where the parties' access to the case documents is exceptionally barred, and whether or not they might have availed themselves of their right to see the documents.

The Public Access to Documents Act does not clearly define the concept *interested party*, but it does appear from the wording of the provision that the term 'interested parties' includes not only those who are legally affected by a decision but others too. According to the preparatory work, the Act and subsequent practice, the main idea is that interested parties are those persons whose interest in the outcome of a case is such as to give them the right of appeal; on the delimitation of this group of persons, see 1.7.3 below. In the field of environmental protection this means that, for example, people who live near an enterprise which creates a certain nuisance have the status of interested parties under the Public Access to Documents Act.[65]

The Public Access to Documents Act does not state the extent to which decision-making administrative authorities are to take the initiative *vis-á-vis* the citizen in drawing his attention to the existence of a case or of case documents about which it would be important for him to know. Without such initiative on the part of the authorities, the citizen would in many cases actually be deprived of the opportunity to exercise the rights conferred on him by part 2 of the Public Access to Documents Act. In this connection an obligation to *hear the interested parties* might exist.

Section 68 of the Environmental Protection Act contains a clause on this subject. According to this clause, the administrative authority must, prior to deciding on an order or prohibition, give written notice to the addressee of the decision (i.e. the person who is responsible for the circumstance in question, cf. section 65, subsection 1, of the Environmental Protection Act) of the case and inform him of his right under the Public Access to Documents Act to be heard. Such notification may be dispensed with, however, if an immediate decision is required or if notification must be regarded as obviously unnecessary. The term 'order or prohibition' must be assumed to include all concrete legal directives whether the decision is formally referred to in the Act as an 'order/prohibition' or not. Thus the authorities must give someone who has received permission to use a cesspool an opportunity to make a statement before the above permission is revoked, even though the provision in section 11, subsection 3, of the Act provides for 'revocation' of such permits.[66] It must also presumably be understood that section 68 of the Act applies to decisions as to orders and prohibitions in appeal cases as well as in first instance cases.[67]

Two limitations to the scope of this clause must be emphasised. In the first place, the provision only applies to the addressees of orders and prohibitions. Any third party who might be interested in whether an application for permission to establish polluting activities is granted or refused is not entitled to be informed, even if he or she may be an 'interested party' within the meaning of the Public Access to Documents Act. It appears from the preparatory works of the Act that the legislators hesitated to suggest a rule about consulting third parties as it is difficult to decide which category of persons to consult in each case. This reasoning does not carry conviction so long as parties other than the addressee of the decision have the right to appeal, cf. section 74 of the Act, and since the rule of communication in section 65, subsection 2, of the Act lays down an obligation to inform these other parties of the decision in the first instance.

Secondly, the entitlement to information under section 68 of the Act hardly extends further than mere notification of the existence of the case. Once a party has received information to this effect, it is up to that party to keep himself posted of the development of the case and of the appearance of documents whose contents it is important to know. Nor do ordinary unwritten administrative rules make it a duty for the competent authority to consult a party if, during the case, information unfavourable to the latter becomes available.[68] Requests made by a party at an early stage of the proceedings for an opportunity to see subsequent statements made by others will, however, usually be complied with in practice.

In Denmark, as in other countries in the western world, there has in recent years been considerable interest in arrangements which allow not only the landowners affected, but also other individuals and associations, through a procedure *based on the principle of publicity*, to influence decisions regarding the use of the land.[69] This interest has manifested itself in comprehensive provisions, especially in legislation on regional plans and town planning, regarding the involvement of the public before a final decision is made.[70]

It is remarkable in the light of this that legislators in the field of environmental protection have been very reluctant to involve the public in the process of decision-making. There is in this field—even where the most extensive installations jeopardising the environment are concerned—no parallel to the American 'Environmental Impact Statements' and the procedure attaching to the latter. The need for environmental reports and for letting the public have its say before a decision is made may be deemed in Danish law to be met in part—but only in part—by the publicity procedures of the planning legislation.

As regards two types of decision which, although specific, have a certain general character, special regulations providing for publicity procedure have been laid down by the notices issued in accordance with sections 16 and 20 of the Environmental Protection Act. The two decisions concern, on the one hand, decisions about protected areas around water supply installations (sections 12 and 13 of the Environmental Protection Act) and, on the other, municipal plans for the enlargement of sewage installations and for the construction and extension of public sewage systems which are not part of an approved sewage plan (sections 21 and 22 of the Environmental Protection Act).[71]

The main content of these regulations is that all landowners, users and authorities who are particularly concerned should be invited to a public meeting where there will be an opportunity for discussion on the basis of the application lodged to draw fresh water and the municipality's draft sewage plan, respectively, both of which have been published at least 3 weeks previously. Verbal statements made at the meeting may be supplemented by written observations up to 3 weeks after the meeting has been held. The category of people invited for meetings concerning sewage plans, etc. is somewhat wider than in cases concerning the water supply.[72] Thus, in the former instance the invitation also includes interested local organisations such as business organisations, nature conservancy organisations and anglers' associations, cf. the Sewage Circular of 17 April 1974 from the Ministry of the Environment.

1.7.3 The right of appeal

In Danish law an important means of protecting the position of the individual in relation to administrative decisions is the so-called 'administrative appeal'. This means the right to complain to superior administrative authorities about concrete decisions taken by subordinate authorities. The appeal authority is obliged to deal with the complaint on its merits and is entitled to alter the decision complained of if it is unlawful and even if it is only deemed inexpedient. In such a case the usual practice is to allow the subordinate authority a certain latitude. As already mentioned, administrative appeal is a legal remedy for the individual, but in this context it may be appropriate to emphasise that, especially in recent years, the appeal often serves a wider administrative purpose as well. Among other things, this may manifest itself as a wide circle of people entitled to complain, sometimes including certain public authorities, rules concerning the suspensive effect of complaints and the right of the superior authority to alter the decision to the detriment of the person complaining.[73]

Today there are in most sectors statutory regulations which more closely determine the pattern of the administrative appeal system, cf. parts 11 and 12 of the Environmental Protection Act. Even where no such regulations exist, there is nevertheless a right to appeal to superior authorities against decisions made by subordinate authorities, and in the last resort to the minister concerned. However, decisions made by boards or by municipal authorities can only be appealed against if there is statutory authorisation do so. In the field of environmental protection this means that the decisions of the Environment Agency may be appealed against to the Ministry of the Environment, even in the absence of any express statutory provisions.

According to the Environmental Protection Act, specific decisions will, as a matter of principle, be taken first of all by a municipal council. These decisions may almost without exception be appealed against to the Environment Agency, cf. section 70, subsection 1, of the Act.[74] Decisions taken by the Environment Agency may in turn, to a certain extent, be referred to the Environmental Appeal Board, cf. section 76 of the Act and 1.3 above. Where a decision may not be appealed against to the Environmental Appeal Board, there is (under section 70, subsection 2, of the Act) a limited opportunity for having the case reviewed by the Ministry of the Environment, viz. where 'the decision is of a fundamental character or of major importance for environmental protection'.[75]

Sections 74 and 80 of the Environmental Protection Act enumerate those who are entitled to appeal to the Environment Agency and the Environmental Appeal Board, respectively, and this enumeration must be considered exhaustive. The provisions of sections 74 and 80 of the Act on the right to appeal are identical in all essentials. The question of appeal will therefore be dealt with taking both lists together, taking section 74 of the Act as a basis and indicating the few points on which the provisions of section 80 of the Act differ.

Those entitled to appeal may roughly be divided into three groups: the addressee of the decision; some public authorities; and private individuals other than the addressee of the decision.

It almost goes without saying that the *addressee* of the decision is entitled to appeal and there is no doubt as to who belongs to this category. In practically all situations, the polluter will be the addressee of the authorities' decision.

The Environmental Protection Act also contains provisions concerning the right of appeal of *certain public authorities*. The clauses apply first and foremost to municipal councils. The basic principle of the Act is that the local council and county council concerned and the Greater Copenhagen Council can appeal against decisions made by another council and against decisions made by the Environment Agency if the case affects the municipality or county in question or the metropolitan area. The background to this is partly the need for coordination as, pursuant to other legislation, the municipal councils undertake tasks which are affected by decisions concerning environmental protection, and partly that municipal councils in a more general sense look after the interests of local communities. Furthermore, the medical officer of health concerned has the right to appeal to the Environment Agency whereas he has no right of appeal to the Environmental Appeal Board.[76] The Act also empowers the Minister of the Environment to lay down rules on or to direct that decisions concerning environmental protection may be appealed against by certain authorities in other countries. Such regulations are laid down in Notice No. 487 of 1 October 1976 on the implementation of the Nordic Convention on Environmental Protection. Under section 4 of the Notice specially appointed 'supervisory authorities' in Sweden, Norway and Finland have the right to appeal against decisions concerning activities which lead or may lead to pollution in the country in question.

Whereas the two categories of persons entitled to appeal are precisely defined by the Act, the wording of the clause concerning *other private individuals entitled to appeal* is vague. They are referred to by the Act as 'anybody who must be deemed to have an individual, essential interest in the outcome of the case', which is no more than the usual criterion in legal theory for a right of appeal.[77] This passage understandably leaves considerable doubt as to which private individuals, other than the addressee of the decision, are entitled to appeal. This question frequently arises in practice and the subject has already been dealt with in several literary works.[78] It should also be mentioned that in practice most appeals are lodged by people within this category of persons entitled to appeal. During the period from 1 January 1975 to 1 July 1977 the Environment Agency dealt with a total of 327 appeals, 7 of which were initiated by public authorities, 74 by the addressee of the decision and the remaining 246 by other private persons with the right of appeal—'neighbours' in a wide sense. During the same period, the Environmental Appeal Board dealt with 177 appeals, 11 of which were initiated by public authorities, 46 by the addressee and 120 by other private individuals.[79] For this reason it will be useful to deal with this group in greater detail than is necessary for the other groups of persons entitled to appeal. First the rules concerning the individuals' rights of appeal will be dealt with and then rules governing the right of appeal of associations.

1.7.3.1 INDIVIDUALS WITH THE RIGHT OF APPEAL OTHER THAN THE ADDRESSEE

One common characteristic of this group of persons entitled to appeal is the negative feature that they are not addressees of the decision, which means that it has no legal effect on them. Another characteristic is the positive feature that the outcome of the case has in reality a considerable effect on their situation. For this effect to form the basis of a right of appeal, the interest attaching hereto must be *individual* and *essential*. That the interest must be individual means *prima facie* that the decision in the environmental case in question must affect the individual personally and its effect on his situation must be special compared with the effect on everybody within a large group. That the interest must be 'essential' means that it must be of certain absolute proportions. For all practical purposes, in a further analysis the two elements of the legal criterion become one, however, because the crux in relation to both will be an assessment of the extent of the interest. A study of the field covered by the rule cannot, therefore, be based on this distinction. Instead, the analysis will use the following criteria: the basis of the interest and the nature of the pollution.

The indirect effect of a decision on environmental protection which can form the basis of a right of appeal will practically always be environmental disturbance. The decision may have a potentially negative effect on the environment and those who live in it will have an interest in the prevention of such negative effects. The crucial point will be what may be termed the *polluting effect or nuisance* of the decision. Other types of effect may conceivably form the basis of a right of appeal. This is the case particularly as regards a *competitive effect* or other economic effects—such as the effect of the decision on the security of a mortgagee. As questions concerning a right of appeal based on the indirect economic effects of a decision on persons other than the addressee of the decision have not, to my knowledge, arisen in practice and will arise only in exceptional cases, the following is written purely with regard to the effects of pollution and nuisance.

For a person to be able to appeal, it must be assumed that his interest is based on either *residence, occupation or right to dispose of real property*. The basis may and will more often than not be the fact that the appellant has his residence near the planned project. According to the passage in question, the most important group of persons entitled to appeal is the 'neighbours' of the projected installation. Whether the person in question is an owner–occupier or a tenant is of no importance in principle. Nor is it a deciding factor whether it is a question of permanent habitation or summer residence only. Also persons who are permanently *employed* near the plant in question and who may be exposed to pollution may be entitled to appeal—this applies to owners and employees alike of nearby firms.[80] Finally, the *right to dispose of real property* can form the basis of a right of appeal if the owner of the right is disturbed by pollution. Thus someone who has fishing rights in a stream could appeal against decisions concerning sewage discharge into it.[81]

Other grounds than the three mentioned may only very rarely, if ever, form the basis of a right of appeal. For example, a person will not be entitled to appeal because he is exposed to a nuisance at a place where he is accustomed to go bathing or which he regularly passes. Neither can hotel guests complain about pollution of the surroundings. In such cases the interest in complaint is generally too small.

Not everybody who has an interest in environmental protection based on residence, occupation or property rights will be entitled to appeal. A further condition is that the person concerned may *actually* be exposed to pollution of a certain magnitude. Assessment of the weight of interest in appeal will depend on the *nature of the pollution*. In cases of emission of noise geographical proximity (depending on the amount and character of the noise) will decide whether or not an appeal lies. The same will apply to nuisance caused by smell or dust.[82] As regards air pollution as distinct from nuisance by smell and dust, the delimitation is more difficult. In the first place the effects of such air pollution will be more widespread, and secondly the extent to which the nuisance is felt will often be insignificant. Yet in these cases too the delimitation in practice seems to be based on aspects of proximity similar to those obtaining for noise, smell and dust.[83] This will probably also necessarily be the way in which delimitation must take place if private individuals other than the addressee of the decision are to be entitled to appeal, although many more persons than the immediate neighbours of the polluting plant will often be exposed to just as heavy pollution from the latter without being entitled to appeal on this basis. In administrative practice, persons who live or work more then a few hundred metres from the source of the nuisance will only exceptionally be entitled to appeal. To this extent, in cases concerning noise and air pollution it is possible to use the term 'neighbour complaints' even though the word neighbour is not to be taken in the strict sense. With water pollution, the question will most often be whether the appellant's use of the recipient is essentially disturbed. Such disturbance will usually, but not always, affect only a rather limited circle, but the activities of such persons may well take place far from the point of discharge.[84]

1.7.3.2 ASSOCIATIONS ENTITLED TO APPEAL

Neither in section 74 nor in section 80 is there any mention of associations as entitled to appeal. The preparatory works of these sections, however, show that the general passage 'anybody who has an individual, essential interest in the outcome of the case' was also to a certain extent aimed at giving associations a right of appeal.

By way of introduction, it is observed that if a corporation—a company, association or the like—is itself as a legal entity exposed to pollution, its management will, like individuals, be entitled to appeal. Where a joint-stock company owns a business or a residential property near a projected scheduled plant, the company is entitled to appeal in the same way as individuals who own such property. Furthermore, there is no doubt that an association can act as agent for an individual entitled to appeal and, by authorisation of the latter, lodge a complaint. The problem is whether corporations, in practice associations, can appeal in their own names because the interests of their members are threatened.

According to the explanatory statement accompanying section 74 of the Bill, the decisive point in that respect is 'whether the said association can be said to *represent* a category of persons whose individual essential interests are directly affected by the decision'.[85] As examples of associations entitled to appeal are mentioned houseowners' associations and anglers' associations whereas 'initiative-taking associations and associations influencing public opinion' are not considered entitled to appeal. When estimating whether or not an association can be said to represent appeal interests its object will be of major importance, but the extent to which the individuals affected are members of the association is also important. Nationwide associations will not as a rule be entitled to appeal, as their object must be presumed to be 'the taking of initiatives and influencing public opinion' rather than looking after specific interests affected.

Where members of an association would be entitled to appeal as individuals, an association with the object of representing the environmental interests of those concerned must also be presumed to be able to appeal in its own name. The most difficult problem arises where no member

AUTHORITIES AND POLLUTION CONTROL

has an interest sufficient to give him a right of appeal. In such a case, can the association appeal representing the members' collective rights? The problem will arise, for example, where a major plant may lead to considerable air pollution for a residential area. In this case the individual residents will not, as a rule, be entitled to appeal, cf. 1.7.3.1 above and note 83, but the question is whether a residents' association will be able to appeal against the decision.

Some good arguments may be adduced in favour of associations being entitled to appeal in such cases. If an association is not presumed to have a right of appeal, the situation will often be one in which nobody exposed to pollution is entitled to appeal and the importance of the public authorities entitled to appeal as representatives of general environmental interests should not be overestimated. As regards sewage plans, there is a basis for the right of appeal of associations, particularly in section 8 of the Sewage Notice according to which 'local organisations' shall be summoned to a public meeting in the same manner as a number of persons and authorities whose right of appeal is undoubted.

Administrative practice is not as yet completely clear on this point but so far seems to indicate a certain, though narrow, margin for appeals by associations in cases where none of the members can appeal as an individual and where the association's rights as a corporation are not affected.[86] In this connection special importance must be attached to a decision of March 1980 made by the Environmental Appeal Board. In this case the Danish Society for the Preservation of Natural Amenities and the Danish Anglers' Federation had appealed against two factories being granted permission to discharge sewage into the Little Belt strait. The Anglers' Federation had argued, among other things, that it represented local circles in the area who fished the Belt and who stocked these waters with fry every year. The Appeal Board held, however, that neither of these associations was entitled to appeal. The Appeal Board recognised the fact that the two associations had a considerable idealistic interest in such cases, but also stated that to accept these associations 'as the natural agents of a more or less clearly defined group of persons' and thus as having 'an "individual, essential interest" in the outcome of the case in question goes beyond the bounds of a proper interpretation of the law'.[87]

1.7.3.3 NOTICE OF THE DECISION OF THE LOWER ADMINISTRATIVE AUTHORITY

A precondition of being able to appeal is that one should receive notice of the decision taken by the lower administrative authority. Naturally this precondition is met in respect of the addressees of decisions, but not with the same degree of certainty for others who are entitled to appeal. In this connection it is important that section 65, subsection 2, of the Environmental Protection Act provides that *notice* of municipal decisions shall be given to the medical officer of health concerned and to other interested authorities and private individuals at the same time as the addressees are informed of them. Notice to private individuals may be given by public advertisement. The Ministry of the Environment issued a circular requesting that municipal authorities communicate decisions by advertisement whenever there is any doubt as to the authorities' knowledge of all the parties entitled to appeal.[88] After the parties concerned have been notified of the decision, there is a general time limit of 4 weeks for appeals both to the Environment Agency and to the Environmental Appeal Board. It is expedient for all parties entitled to appeal to be notified at the same time, as otherwise the appeal time limits will start running at different times.[89]

1.7.4 Suspensive effect of appeals

If an appeal is to have the intended effect, it is often important that the matter is left in abeyance in the circumstances that existed before the lower administrative authority made its decision. If an order can be enforced before the appeal authority finishes its procedure in the case, the interest in lodging an appeal will often be limited and the appeal authority's handling of an appeal against a permit will be considerably prejudiced if the permit can be used irrespective of the appeal lodged. On the other hand, important interests also attach to appeals not leading to postponement of the enforcement of orders/injunctions or of the opportunities of using permits.

In sections 72 and 81, subsection 2, the Environmental Protection Act takes account of these rival interests. By virtue of these provisions, appeals to both the Environment Agency and the Environmental Appeal Board against *orders and prohibitions* have a suspensive effect unless the appeal authority decides otherwise in the specific case. Furthermore, a municipal council can decide that orders and prohibitions shall be complied with irrespective of an appeal to the Environment Agency if particular reasons—such as a real health hazard—call for it.

With regard to *permits*, section 72, subsection 3, of the Act provides that if their use involves construction work, such work shall not be started until the period allowed for appeals to the Environment Agency has expired. If an appeal against the permit has been lodged in due time, the construction work must not commence until the appeal authority's decision is available. This means that appeals against certain permits also have a suspensive effect, which is interesting because this is contrary to the general unwritten rule of Danish administrative law. In practice, it was assumed that there were no exceptions to this rule, which might have adverse consequences.[90] By virtue of an amendment to the Act in 1978, the appeal authority is now entitled while dealing with the appeal to allow construction work to be commenced when special circumstances call for it and it is otherwise considered unobjectionable.[91] The Minister of the Environment too has been empowered generally to exempt certain minor construction works.[92] Appeals to the Environmental Appeal Board against permits granted or confirmed by the Environment Agency, however, never have a suspensive effect.

1.7.5 The right of action (*locus standi*)

There are various ways in which questions of environmental protection can be brought before a court. In the first place, it may be done during a criminal case, in which a citizen is charged with having violated the legislation on environmental protection, because he will be convicted only if the court finds that the administrative decision is lawful. Secondly, the citizen may himself bring a civil action before the courts concerning a question of pollution and there are several other—albeit in practice less important—ways in which similar matters may be brought before the courts.

The present subsection deals solely with the citizens' opportunities for bringing actions to have decisions made by administrative authorities on environmental protection issues set aside. A general right to have such questions tried by a court is provided by section 63 of the Danish Constitution empowering the courts to decide whether the administrative authorities have acted *ultra vires*. According to legal usage, this means that the courts may overrule administrative decisions if, for technical reasons or for reasons of substantive law, the latter may be deemed unlawful. It is, however, the general opinion that the courts cannot examine the 'discretion' of the administration and they are not normally entitled in conjunction with overruling a decision to make a fresh decision with a different content. Unless provided for by the individual statute, legal action against an administrative authority for the purpose of having its decision overruled has no suspensive effect, whether orders/prohibitions or permits are involved.

Legal action to have an administrative decision overruled is not conditional upon the plaintiff having exhausted the possibilities of complaint to superior administrative authorities, but in practice most plaintiffs normally first make use of the generally quicker, cheaper and more comprehensive administrative appeal.[93]

Not everybody is entitled to bring actions in the courts contesting an administrative decision. It is a condition that the person in question is directly affected by the matter in dispute. The delimitation of the circle of persons entitled to sue follows the principles applying to the delimitation of the group entitled to appeal (in this connection see 1.7.3 above). The basic criterion in both cases is that the person concerned must have an individual, essential interest in the outcome of the case and this standard must probably also be complied with for the right to bring actions and the right of appeal.[94] However, particular rules of law concerning the right to bring actions and the right of appeal, respectively, cannot automatically be applied to the other faculty. Thus as regards decisions under the Environmental Act, medical officers of health will hardly be entitled to bring actions although they are entitled to appeal to the Environment Agency. The provisions of the Act on the right of the medical officer to appeal are considered to be specific. Otherwise those entitled to appeal under the Environmental Protection Act may be assumed to be entitled to bring actions as well.

As mentioned, the administrative appeal only relates to concrete administrative decisions.[95] However, actions may be brought in the courts to have one or more provisions of a regulation issued by the administrative authorities set aside in a specific case as unlawful, more often than not because the regulation lacks legal authority as regards the points mentioned. Such a question will often arise during a criminal case where a person is charged with infringement of a regulation. It may also be raised, however, in an action for recognition and, if so, the criterion 'individual and essential interest' will also be relevant in determining whether the plaintiff may be deemed to have the right of action.

As mentioned in 1.2 above, there are few published judgments on the contestation of administrative decisions in the field of environmental protection and, with the exception of criminal cases, most judgments have concerned questions of technical error. Despite the stepping up of public control of pollution resulting from the Environmental Protection Act and other new acts on environmental protection, there is no indication that the control exercised by the courts will be any greater than under the previous legislation. A contributory cause is probably the setting up of the Environmental Appeal Board with its judicial character.

1.7.6 Actions for compensation

The ordinary appeal authorities in the field of environmental protection cannot award a citizen compensation. If a citizen wants compensation, he must either bring an action in the courts or possibly apply to certain administrative bodies, in particular the Agricultural Land Tribunals. The rules governing compensation are very comprehensive and a detailed account of them could not be contained within the framework of this book, but a few features should be mentioned.

The basis of a claim for compensation may be found either in rules governing *damages* or in rules governing expropriation. As regards compensation under the law relating to adjoining properties, see 1.7.7 below. The rules as to when damages are payable are mainly unwritten in Danish law and based on legal usage in concrete cases.

A claim for damages may be raised firstly on the grounds that loss has been suffered because of a *wrongful administrative decision*. Such a claim will most often be raised by persons who operate or wish to operate a polluting plant and who have, without justification, received a concrete order/prohibition or a refusal. In many cases the mistake may be rectified in that the decision is overruled by an appeal authority or a court and another, correct, decision is made. In some cases, however, the wrong decision leads to financial loss. According to current opinion, compensation should be paid from public funds only if a loss is attributable to fault or negligence (*culpa*) on the part of administrative personnel. The fact that a decision with unlawful content has been made cannot in itself afford grounds for a claim for compensation if no-one in the administration can be blamed for the illegality. In recent case law, particularly in the field of building legislation, there seems to be a trend towards objectification of the responsibility in the sense that the administration must unconditionally bear the cost of losses arising from decisions based on a wrong conception of law.[96] It should be added that inadequate control and failure to intervene on the part of public authorities may render the public purse liable for damages. Definite *culpa* will usually be required in such cases, however.[97]

Secondly, claims for damages may be raised *for losses due to* a person or his property having been exposed to *pollution*. Whereas in the first group claims are usually raised by 'polluters' against a municipality or the state, these claims are raised by individuals who have suffered from the consequences of pollution and the claims are made against the polluter, which may be either a private firm or a public authority. Claims of the type in question will arise especially where isolated arrangements lead to loss owing to pollution of groundwater, watercourses, lakes and the sea.

The basis of responsibility follows the general unwritten civil-law rules on liability in damages. Under these rules liability in damages usually presupposes that the person against whom the claim is raised has acted unwarrantably and that his conduct may be considered intentional or negligent (the *culpa* rule). For many years, however, there has been a trend in legal usage towards making the traditional *culpa* liability more stringent even though the courts have only in a few exceptional cases put forward a purely objective basis as grounds for responsibility. This trend has also been seen as regards damage by pollution. This may be illustrated by a judgment in 1969 (U 1969.923) where the Court of Appeal held that the owner of an oil tank from which oil had leaked out to pollute a water supply installation was liable for damages. Certainly the grounds of the judgment may be referred to the *culpa* principle, but the duty of care imposed on the owner, which the latter did not observe, must be considered extremely severe.[98]

An important question of general interest is what significance can be attached to preventive regulations under public law in relation to the law of torts. Certainly it cannot always be concluded in cases of infringement of such regulations that, for the purpose of the law of torts, the person concerned has acted unwarrantably. In the field of environmental protection, however, the preventive regulations very often express a standard duty of care established in order to provide protection against damage caused by pollution and disregard of the regulations will be deemed to be unwarrantable conduct. In this connection it should be mentioned that the Watercourses Act,[99] which applied until the Environmental Protection Act took effect on 1 October 1974, imposed on owners of *unlawful* sewage installations an unconditional liability for damage caused through the existence and use of the installation. Damage caused by a *lawful* installation, however, was only to be compensated for if it was due to the system not being kept in proper repair, not being attended to or to its being overloaded or used for purposes significantly different from those approved by the Agricultural Land Tribunal. In the environmental reform of 1973 these provisions were not included in the Environmental Protection Act. The reason given was that the provisions of the Watercourses Act were not found to differ significantly from the general rules of Danish law on damages.[100]

In two important fields of legislation special rules on compensation have been laid down. In the first place, they concern damage caused by nuclear plants,[101] (see Chapter 8) and, secondly, they concern damage caused by oil at sea[102] (for further details see Chapter 5). In both cases rules have been laid down imposing strict liability.

Furthermore, *expropriation* may afford grounds for claims for compensation. According to section 73 of the Danish Constitution, a person shall not be forced to give up his property without compensation in full. This provides for protection not only against actual surrender of land[103] but also against certain other qualified encroachments on property. In regard to environmental protection, this rule is of comparatively little significance as the bulk of restrictions on the owners' use of their property for purposes of environmental protection must be considered as more precise interpretations of the limits of ownership rights and not partial expropriation. This applies both to limitations of the right of future disposal and to interference with current activities. In certain respects decisions on environmental protection may be in the nature of expropriation, however. Decisions made under sections 13 and 14 of the Environmental Protection Act providing for the protection of water supply installations may thus be expropriatory if the owners of the areas to which the protective measures apply are compelled to alter or discontinue an existing, lawful activity.[104] In such cases compensation is payable and, failing agreement, the amount shall be decided by the Agricultural Land Tribunal.

During the reading in the Folketing in 1973 of the Environmental Protection Bill the opposition expressed the wish that rules be laid down concerning compensation to lawfully established firms for additional costs incurred as a consequence of demands in pursuance of the Environmental Protection Act. The Government refused to introduce a general rule on compensation. Subsequently the Government set up a committee commissioned to work out a detailed report on the issue. In its report 'Environmental Legislation and Compensation', issued in October 1976, the committee strongly advised against the adoption of special rules on compensation. Such rules would, in the opinion of the committee, involve a dangerous breach with legislative practice concerning payments by the public authorities as a consequence of regulation of land use.

Where a major industrial plant is approved even though it leads to heavy pollution or danger of it, the neighbours will probably be able to put forward claims for compensation based on considerations of expropriation law.[105] There is no legislation concerning such situations, but it should be noted that in one single case, in pursuance of the old legislation, the authorities approved the construction of an oil refinery on condition, among other things, that the company should take over the property within a closely-defined area, against compensation, if the owners were exposed to substantial inconvenience and requested that their property be taken over.[106] In my opinion, there is nothing to prevent a similar condition being laid down in connection with approval, under the Environmental Protection Act, of a polluting plant if particular circumstances actually exist which tell in favour of it. The Environment Agency however, is doubtful whether sufficient legal authority exists for the application of such take-over conditions.

1.7.7 The general law of adjoining properties

In this connection mention will finally be made of the complex of mainly unwritten law known as *den almindelige naboret* ('the general law of adjoining properties'). The main principle is that land may not be used in such a way that the neighbours are exposed to essential nuisances of a permanent nature. When establishing the norm for permissible nuisances, certain factors are taken into consideration, e.g. the degree, duration and frequency of the nuisance, the nature of the place, and which nuisances may be anticipated elsewhere in similar circumstances. If the nuisances are considered excessive, the court may issue an injunction to restrain the activity in question and the neighbour may also be granted compensation. Here too the liability in damages is, in principle, based on *culpa*, but there are various extensions of the basis for liability. Among other things, compensation is undoubtedly payable to the neighbour if operations creating a serious nuisance are permitted to continue for social reasons.[107]

The general law of adjoining properties applies in part to the same field as the rules of environmental protection in public law which are the subject of this account (e.g. in relation to nuisances such as smoke, noise and vibration), and may then be applied concurrently with these.[108] The existence of more comprehensive rules of public law which may be applied more easily does, however, somewhat restrict the practical significance of the law of adjoining properties.

Notes

1. Cf. Sections 3, 15 and 28–32 of the Constitution.
2. Cf. sections 67–80 of the Constitution.
3. Cf. section 82 of the Constitution.
4. This is disregarding the fact that fairly extensive 'home rule' may be granted to areas like the Faroe Islands and Greenland, areas with a character all their own as regards the population as well as geographically and culturally.
5. Cf. section 63 of the Constitution.
6. Cf. especially Bent Christensen, *Forvaltningsret Apparatet*, 1976 (stencilled) and Lis Sejr in *Forvaltningsret, Almindelige emner*, 1979, p. 17 ff.
7. Cf. section 3, clause 2, of the Constitution.
8. Cf. reports Nos. 301/1962, 320/1962, 629/1971 and 743/1975.
9. The type of directorate now prevailing is based especially on report No. 629/1971 on central administrative organisation.

AUTHORITIES AND POLLUTION CONTROL

10. For further details, cf. Act No. 331 of 13 June 1973.
11. For further information on central administration of municipalities, see Bent Christensen, *Dansk Miljøret*, vol. 1, 1978, p. 98 ff.; Erik Harder, *Dansk Kommunalforvaltning, Opgaver og tilsyn*, 1980, p. 215 ff.; and Claus Haagen Jensen, *Kommunalret*, 1972, p. 67 ff.
12. Concerning Danish law, among others see Preben Stuer Lauridsen, *Retslæren*, 1977, p. 265 ff.
13. Claus Haagen Jensen, *Forvaltningsret, Almindelige emner*, 1979, p. 89 ff. contains an account of matters relating to sources of law within administrative law to which the rules concerning the protection of the environment must be deemed to belong.
14. As amended by Act No. 288 of 26 June 1975, Act No. 107 of 29 March 1978, Act No. 220 of 24 May 1978, Act No. 304 of 8 June 1978 and Act No. 212 of 23 May 1979.
15. See also section 5 of the Act which is utilised to only a limited extent in the Environmental Protection Regulations.
16. It should be added that besides decisions by the Environmental Appeal Board, KFE contains other types of decisions concerning real property, e.g. relating to nature conservation and expropriation.
17. Surveys of the information and investigation activities of the Environment Agency appear in *Nyt fra miljøstyrelsen*, Nos. 11/1977, 2/1978, 1/1979 and 4/1979.
18. A more detailed account of administrative organisation in the field of environmental protection is given by Claus Haagen Jensen, in *Dansk Miljøret*, vol. 3, 1977, p. 30 ff.
19. Cf. sections 4, 6 and 10, section 11, subsection 2, sections 16 and 20, section 35, subsection 2, sections 36, 43, 56, 62 and 64, section 74, subsection 3, section 76, subsection 3, section 77, subsection 3, and section 80, subsection 2, of the Act.
20. Cf. Act No. 267 of 8 June 1977 on the content of Lead, etc. in Petrol and Notice No. 354 of 21 June 1977.
21. In this connection refer to *Folketingstidende* 1972–73, appendix A, col. 3944 and appendix B, col. 2196.
22. Cf., for example, section 11 of the Environmental Protection Act.
23. Cf., for example, sections 38, 47, 57 and section 70, subsections 2 and 3, of the Environmental Protection Act.
24. Cf. section 45, subsections 3 and 4, and section 46 of the Environmental Protection Act.
25. For further information on the delegation of the Ministry's powers and their limitations, see Claus Haagen Jensen, *op.cit.*, pp. 38–41.
26. Cf. section 70, subsection 1, of the Environmental Protection Act.
27. Cf. section 70, subsection 2, in connection with section 76 of the Act.
28. Cf. section 70, subsection 3, of the Act *per contra*.
29. For further information, cf. Johan Garde, *Ankenævn i miljøretten, Juristen og Økonomen*, 1989, p. 55.
30. A similar involvement of the Environmental Appeal Board was suggested in a government bill of 14 December 1979 for the amendment of the Urban and Rural Areas Act.
31. Cf. Notice No. 372 of 16 July 1975 issued by the Ministry of the Environment as amended by Notices No. 69 of 1 March 1977, No. 492 of 20 September 1978 and No. 17 of 11 January 1980.
32. See expressly section 75, subsection 2, of the Environmental Protection Act.
33. Cf. Pouel Pedersen in *Nordisk administrativt Tidsskrift*, 1973, p. 104 ff. and the first edition of this book, p. 15 ff.
34. Cf. letter of 4 August 1975, for example, from the Environmental Appeal Board, KFE 1976, p.3.
34a. For further information concerning the system of environmental subsidies, see Notice No. 358 of 11 July 1978, Notice No. 205 of 3 May 1978, *Nyt fra miljøstyrelsen*

NOTES

No. 4/1978 and, regarding the system of subsidies until 1 June 1978, Claus Haagen Jensen, in *Dansk Miljøret*, vol. 3, 1977, p. 23 ff.

34b. Notice No. 292 of 13 May 1976 on the rules of procedure for the Environmental Credit Council issued by the Ministry of the Environment.
35. On various ways of organising committees for environmental protection tasks, see circular of 14 February 1974 from the Danish Association of Municipalities and circular No. 55 of 22 March 1974 from the Ministry of the Interior.
36. Cf. section 39 of the Environmental Protection Act and Notice No. 176 of 29 March 1974, as amended by Notice No. 290 of 28 June 1978.
37. Cf. section 11 of the Environmental Protection Act and Notices Nos. 172 and 173 of 29 March 1974.
38. Cf. section 18 of the Environmental Protection Act and Notice No. 174 of 29 March 1974.
39. Cf. section 4 of the Environmental Protection Act and Notice No. 170 of 29 March 1974.
40. In this connection, cf. sections 18, 25 and 39 of the Environmental Protection Act and Notices Nos. 174 and 176 of 29 March 1974.
41. Cf. section 39, subsection 3, section 44, subsection 5 and section 50 of the Environmental Protection Act.
42. For further information, see sections 61 and 62 as amended by Act No. 288 of 26 June 1975, circular of 22 February 1977 from the Ministry of the Environment, *Vejledning fra miljøstyrelsen* No. 2/1974 and Claus Tønnesen, in *Dansk Miljøret*, vol. 3, 1977, p. 45 ff.
43. Cf. section 59 of the Environmental Protection Act with reference to the Foods Act, Act No. 310 of 6 June 1973, sections 51 and 52.
44. In this connection see section 34, section 39, subsections 2 and 3, section 44, subsection 5, and section 50, subsection 2, and, in connection with subsection 2 of section 39, Notice No. 176 of 29 March 1974, section 4, subsection 2.
45. Cf. sections 4, 6–8 and section 11, subsection 2, and sections 20 and 62 of the Act.
46. Cf. sections 16, 20 and 43 of the Act.
47. Cf. section 9 of the Act on the approval of models of measuring instruments, etc. and section 56 of the Act on supervisory and control operations in general.
48. Cf. section 70, subsection 1, of the Environmental Protection Act.
49. For a detailed account of central administration in the field of environmental protection, see Claus Haagen Jensen, in *Dansk Miljøret*, vol. 3, p. 41 ff.
50. For further details concerning the composition and mode of operation of the Agricultural Land Tribunals, refer to Notice No. 133 of 6 March 1970 on the Agricultural Land Tribunals Act. A list of existing Agricultural Land Tribunals is published in *Nyt fra miljøstyrelsen* No. 5/1978.
51. Cf. section 14, subsection 2, and section 32, subsection 2, of the Environmental Protection Act.
52. Cf. sections 92 and 93 of the Environmental Protection Act and Notice No. 403 of 13 August 1976.
53. Cf. section 59 of the Environmental Protection Act.
54. For further information on the medical officer agencies, see Act No. 381 of 13 June 1973.
55. In this connection, cf. section 51, subsection 2, of the Environmental Protection Act.
56. Cf. Notice No. 124 of 7 April 1967. When the Marine Environment Act of 7 April 1980 comes into force, the Oil Pollution Board will be abolished.
56a. See in this connection also Section 44 of the 'Chemicals Act', No. 212 of 23 May 1979, according to which the Minister of the Environment sets up expert boards to assist the authorities on issues concerning chemical subtances and products. The Minister is further authorised to lay down rules on the extent to which the opinions of such boards should be requested before decisions are made.

57. On the concept of 'grass-roots' movements and their relations with interest organisations, readers are referred to Poul Erik Mourtisen, in *Politica* No. 4, 1979.
58. No. 174 of 29 March 1974.
59. Cf. Folketingstidende, 1972–73, appendix A, col. 3943.
60. Act No. 280 of 10 June 1970.
61. In a report on a revision of the Public Access to Documents Act No. 857/1978 this limitation of the access of the public to documents has been preserved in principle, but it is suggested that the Minister of Justice should be entitled to demand that specified authorities shall draw up lists of mail received, see 18 ff. of the report.
62. For details of the general access of the public to documents, see in particular, N. Eilschou Holm, *The Public Access to Documents Act*, 1970, pp. 15–122 and the report mentioned in note 61.
63. On the previous state of law, see Claus Haagen Jensen, in particular, in *Dansk Miljøret*, vol. 3, 1977, pp. 128–131, in which he argues in favour of a restrictive interpretation of section 85 of the Environmental Protection Act, and *FOB* 1977, 362.
64. The report mentioned in note 61 suggests that no rule of secrecy shall in itself entail limitation of the access to public authority documents, see p. 30 ff. of the report.
65. On part 2 of the Public Access to Documents Act, readers are referred, moreover, to N. Elischou Holm, *The Public Access to Documents Act*, 1970, pp. 123–59.
66. For further information, see Claus Haagen Jensen, *op.cit.*, p. 122 ff. The Environment Agency may adopt a different view as regards this issue: in this connection, see *Nyt fra miljøstyrelsen*, No. 5/1976, p. 23, concerning a decision in pursuance of section 31 of the Act, on expropriation.
67. Niels Borre, in *Miljøbeskyttelsesloven*, 1973, p. 164, does not think that section 68 should be applied to decisions made by appeal authorities. This restrictive interpretation, however, has no support in the Act or the preparatory works of it. On the contrary, sections 65 and 66 of the Act, which are expressly limited to decisions made by lower authorities, seem to support the view that in the absence of a specific reservation, section 68 of the Act shall cover all decisions concerning orders and prohibitions.
68. Cf. C. A. Nørgaard, *Forvaltningsret, Sagsbehandling*, 1972, pp. 98–104. The Ombudsman has, however, in recent years presupposed the existence of a certain duty to allow the party in question to be heard even though it does not follow from express provisions, see *FOB* 1972 40, 1973 86 and 1975 610, for example.
69. Cf. Bent Christensen in *TfR* 1976, p. 39 ff.
70. The Acts on Regional and Town Planning date from 1973 and 1975, respectively.
71. See Notice No. 180 of 29 March 1974, sections 7–11 and Notice No. 174 of 29 March 1974, sections 8–12 and 17, respectively.
72. Cf. Notice No. 174 of 29 March 1974, section 8, in connection with Notice No. 180 of the same date, section 7.
73. On the subject of administrative appeal generally, see especially Poul Andersen, *Dansk Forvaltningsret*, 5th ed. 1965, pp. 522–41; Haagen Jensen and Nørgaard, *Administration og Borger*, 1973, pp. 179–96; Ellen Margrethe Basse, in *Forvaltningsret, Almindelige emner*, 1979, pp. 289–333; and Report No. 657/1972.
74. Municipal decisions concerning the distribution of costs of sewage systems where the appeal concerns 'legal issues' only are exempted, cf. section 27 subsection 8, of the Environmental Protection Act as amended by Act No. 107 of 29 March 1978. The amendment relating to section 27 has not yet come into force, however.
75. For detailed information on the administrative appeal system within the field of the Environmental Protection Act, see Claus Haagen Jensen, in *Dansk Miljøret*, vol. 3, 1977, pp. 132–40.
76. According to the original wording of the Environmental Protection Act, the town planning committee and the conservancy planning committee also could appeal against municipal decisions to the Environment Agency. The aim of this provision

was to give the bodies mentioned a right of appeal for the purpose of protecting general public interests within their spheres of competence and the medical officer of health has a right of appeal in his field. Amendments of 1975 and 1978 abolished the right of appeal of both committees in conjunction with their dissolution.
77. Cf. Poul Andersen, *op.cit.*, p. 531 ff.
78. Cf. Johan Garde, *Juristen og Økonomen*, 1978, p. 489 ff.; Ellen Margrethe Basse, *Juristen og Økonomen*, 1979, p. 486 ff. (particularly on the right of appeal of 'grass-roots organisations'); and Claus Haagen Jensen, *op.cit.*, p. 142 ff.
79. This information is taken from Ellen Margrethe Basse, 'Etude statistique concernant les procédures contentieuses en matière de l'environnement dans les Etats membres des Communautés Européennes', *Etude* No. 2, pp. 49 and 53.
80. Workers who expect to be employed by the projected plant will not be entitled to appeal, however, cf. letter of 7 February 1977, *KFE* 1977, p. 74. Such workers are only protected by the Worker Protection Act.
81. Cf. the Environment Agency's letter of 15 September 1976 (unprinted).
82. Cf. decision of the Environmental Appeal Board of 18 December 1978, *KFE* 1979, p. 76, in which people living some 400 m from a poultry farm with 30,000 chickens were deemed not to be affected by smell from the farm to such an extent that they would have an individual essential interest in the outcome of the case. This decision may be compared with the letter of 6 December 1977, *KFE* 1978, p. 65, in which houseowners at a distance of at least 320 m were deemed entitled to appeal in a case concerning approval of a drug factory.
83. Cf. the Environmental Appeal Board's letter of 7 February 1977, *KFE* 1977, p. 74, in which an appeal from a person living some 1·5 km from the plant was rejected. The appeal concerned the expected discharge of some metals, the effects of which would be considerable over a large area.
84. Cf. the Environment Agency's letter of 15 September 1976 (unprinted) concerning an appeal from the owner of fishing rights some 20 km from the place of discharge.
85. Cf. *Folketingstidende* 1972–73, appendix A, col. 4013 ff.
86. As an example of a case in which the rights of the corporation were affected by a decision, there may be mentioned the letter of 3 July 1975, *KFE* 1975, p. 170, from the Environmental Appeal Board.
87. Letter of 24 March 1980 from the Environmental Appeal Board, published in *KFE* 1981, p. 95. For more about the problem, see Claus Haagen Jensen, *op.cit.*, p. 146 ff. and Ellen Margrethe Basse, *Juristen og Økonomen*, 1979, p. 485 ff., and concerning the specific decision of 24 March 1980, the ombudsman in *FOB* 1980, p. 501.
88. Circular of 15 May 1974, part 4, item 3, from the Ministry of the Environment.
89. For further information on time limits for appeals and their significance, see Claus Haagen Jensen, *op.cit.*, p. 149, and Jørgen Hansen and Jørgen Bjerring in *U* 1974, B, p. 214 ff. and 389 ff., respectively.
90. Cf. *Nyt fra miljøstyrelsen* No. 5/1976, p. 11. This assumption was criticised by Claus Haagen Jensen, *op.cit.*, p. 151 ff.
91. Act No. 107 of 29 March 1978, according to which exemption rules, subsections 4–6, were added to section 72 of the Environmental Protection Act.
92. This was effected by Notice No. 505 of 10 October 1978.
93. As regards taking administrative decisions to court, readers are referred especially to Poul Andersen, *Dansk Forvaltningsret*, 5th ed., 1965, pp. 569–650; Bent Christensen, *Højesteret* 1661–1961, pp. 354–401; Ole Krarup, *Øvrighedsmyndighedens grænser*, 1969; and Jørgen Mathiassen, in *Forvaltningsret, Almindelige emner*, 1979, pp. 203–88.
94. In legal theory opinions differ on this subject, cf. Ellen Margrethe Basse, *Juristen og Økonomen*, 1979, p. 492 ff. with references. A judgment by the *Landsretten* ('High Court of Justice') of 21 October 1977 possibly reflects the position that the right to bring actions is more limited than the right to appeal. In this case a very loose-knit

grass-roots movement was considered entitled to appeal against a decision made in pursuance of the Municipal Planning Act which does not contain special provisions concerning the right to appeal but the organisation was not considered entitled to bring a legal action. This might be due to procedural law requirements other than the rules concerning the right to bring actions. For further reference on the judgment, see Ellen Margrethe Basse, *op.cit.*, p. 488 ff.
95. It must be presumed that general decisions made by the local councils in pursuance of the environmental protection regulations may also be appealed against under the general provisions of the Environmental Protection Act, cf. Claus Haagen Jensen, *op.cit.*, p. 134 ff. and P. Spleth, in *U* 1979 B 39.
96. On liability in damages under public law, readers are referred especially to Report No. 214/1959 and to Orla Friis Jensen in *Forvaltningsret, Almindelige emner*, 1979, pp. 371–406.
97. In this connection, cf. *U* 1978, 585 ØL.
98. For further details on this judgment, see Stig Jørgensen, in *U* 1970 B 192 ff. and K. Skovgaard-Sørensen, in *U* 1971 B 36. On general liability in damages under civil law, readers are referred especially to Stig Jørgensen and Jørgen Nørgaard, *Erstatningsret*, 1976 and Anders Vinding Druse, *Erstatningsretten*, 3rd ed., 1976. On general liability in damages for damage caused by pollution, refer to Jens Christensen in *Juristen og Økonomen*, 1977, pp. 453–65.
99. Statutory Notice No. 132 of 6 March 1970, section 72.
100. Cf. *Folketingstidende* 1972–73, appendix A, col. 4022.
101. Act No. 332 of 19 June 1974 on damages in case of nuclear damage.
102. Part 12 of the Merchant Shipping Act, as amended by Act No. 227 of 24 April 1974.
103. An express provision on the subject can be found in section 31 of the Environmental Protection Act which authorises compulsory purchase for the purpose of establishing a sewage plant.
104. For further details cf. Claus Haagen Jensen, *op.cit.*, p. 158 ff. The predominant view of Bernhard Gomard on damages under section 13 and 14 of the Environmental Protection Act seems to be that such compensation is equitable, not compensation for expropriation, cf. *U* 1978 B 61 ff.
105. For further details cf. Claus Haagen Jensen, *op.cit.*, p. 161 ff.
106. Letter from the Ministry of the Interior dated 31 May 1960.
107. On the general law of adjoining properties, readers are referred especially to W. E. von Eyben, in *Dansk Miljøret*, vol. 4, pp. 281–317, and to Knud Illum, *Dansk Tingsret*, 3rd ed., by Vagn Carstensen, pp. 99–118.
108. Cf. Claus Haagen Jensen, *op.cit.*, p. 13 ff.

2
Air

2.1 STATIONARY SOURCES OF AIR POLLUTION

2.1.1 Control by the general planning legislation

The most important rules governing the location of plants which may cause air pollution—as well as most other forms of pollution—are to be found in the Environmental Protection Act. The control which is exercised on this basis over the location of polluting plants, however, takes place within a framework which is laid out in the general planning legislation. This framework, which is an important element in the limitation of nuisances as a consequence of air pollution, among other things, will be discussed in this subsection whereas the control in pursuance of the Environmental Protection Act over the location of plants causing air pollution is dealt with in 2.1.2.

Physical planning in Denmark is mainly governed by three Acts: the National and Regional Planning Act 1973,[1] the Urban and Rural Zones Act 1969,[2] and the Municipal Planning Act 1975.[3] These three Acts are the cornerstones of the so-called 'planning Acts reform', which brought about a radical three-stage reform of the general rules on planning and the use of the country's physical resources. The sequence of Acts—which is obviously not chronological—indicates partly a transfer from large to small planning areas, and partly an increase in the degree of detail in the plans.

So far there are no actual *national plans*, but since 1961 preparatory work has been carried out with a view to this sort of planning. Under the National and Regional Planning Act, the Minister of the Environment is bound to see to it that comprehensive national physical planning is carried out. After consultation with other ministers whose departmental interests are particularly affected, he can, as part of the national planning, lay down directives which will form the basis of regional planning. So far few such directives have been issued. As the most important example there may be mentioned a circular concerning the reservation of areas for the main distribution network for natural gas in Denmark.[4] Furthermore, it should be noted that the Minister of the Environment is under obligation to give an annual account of the national planning work to the committee on physical planning set up by the Folketing. This report is usually made the subject of a debate in the Folketing and although it does not contain binding norms its contents are nevertheless of some importance to the planning work at lower administrative levels.

According to the National and Regional Planning Act a *regional plan* is to be worked out for each county outside the metropolitan area (on the latter, see below). This plan is worked out by the county council and approved by the Minister of the Environment. Among other things the plan must include guidelines for the distribution of future urban growth in the various parts of the county, the extent and location of the most important centres, major traffic schemes and large public institutions, and the location of activities, etc. for which there are special requirements as regards location with a view to the prevention of pollution. The guidelines of the approved regional plan must be followed by the county council and the local municipal councils in their planning (e.g. municipal planning) and public works as well as in their exercise of powers under legislation (e.g. the Environmental Protection Act). On the other hand, the regional plan is not directly binding on the citizens but, because it binds the authorities, it will also indirectly affect individuals.

The development of a regional plan is a comprehensive and complicated procedure involving the public at two different stages. The working out of regional plans started when the Act came into effect on 1 April, 1974. At the end of 1979 final draft regional plans were available from all county councils. The publication of these drafts together with a considerable amount of supplementary material is followed by a four-month period during which objections may be raised. When the county council has forwarded the objections and comments received together with its own evaluation of the latter to the Minister of the Environment, he can consider approval of the draft regional plans with amendments, if necessary. Approved regional plans are expected to be available for all counties in 1981.[5] Supplements to the regional plan can be carried through subsequently—maybe on the basis of an order from the Minister of the Environment—following a procedure which may be considerably less comprehensive than the procedure for the basic regional plan itself, for which there was no parallel in the previous legislation. This means that the regional plans can be amended within a much shorter period than that required to carry through the primary regional plan.

The regional planning is in the nature of 'comprehensive physical planning'. This means that the regional plan must contain a planning summary of the use of the physical resources within the various sectors. In the field of environmental protection the county council must, in pursuance of section 61 of the Environmental Protection Act,[6a] prepare on the one hand, a survey of sources of pollution in the county and, on the other, plans for the future location of industry for which there are special siting requirements for the purpose of preventing pollution. Guidance as to which types of industry are covered by the provision and as to requirements regarding location is to be found in a circular of 1977 from the Ministry of the Environment.[6b] These plans are to be included in the regional plan and, when it is approved, become binding on the authorities, see section 62 of the Environmental Protection Act.

Until that time, pursuant to section 61 of the Environmental Protection Act, the county council's plans serve only as a guide. The duty of the county council to work out plans for siting industry for which special requirements as to location must be imposed is construed to mean that before working out these plans the county council shall, as far as possible, set a target for environmental quality in the area. For this reason planning under section 61 of the Environmental Protection Act is usually described as 'environmental quality planning'.[7] So far such recipient-quality requirements have been laid down solely for surface waters (watercourses, lakes and the sea).[8]

The comparatively rough outline provided by a regional plan is supplemented by planning regulations under the Municipal Planning Act which took effect on 1 February 1977. This takes place through implementation of a *municipal plan* and local plans. Like the regional plan, the municipal plan is a new element in Danish planning legislation and, like the former, it has no directly binding effect on the citizens. The municipal plan is comprised partly of general planning of the 'main structure' for area use throughout the municipality, and partly of frameworks for local plans. The plan can be adopted by the local council after a formative procedure which involves the public at two stages among other things. The principal rule is that other authorities cannot stop a municipal plan which is in accordance with the regional plan and national planning directives, if any. There are certain exceptions, however. It should be mentioned, in particular, that the Minister of the Environment has discretionary powers to require that a concrete draft municipal plan shall not have legal effect unless approved by him. The Municipal Planning Act provides that all municipalities shall set up municipal plans. The obligation to do so does not take effect, however, until a regional plan has been approved including the municipality and drawn up in such a way that draft municipal plans are available not later than two years after the approval of the regional plan covering the municipality. The effect of this has been that work on municipal plans is only just starting and that they will probably not be available until 1982–83. Until municipal plans have been executed, the Minister of the Environment can approve a temporary framework for the contents of local plans, the so-called 'section-15 framework'.

Local plans concern only limited parts of the area of a municipality and have a direct legal effect for private individuals. Local plans may affect a number of different things, for further details see section 18 of the Act. Among other things the plan may contain provisions on the distribution of different types of building (e.g. residential accommodation and office and industrial buildings), on the use of individual premises, on the scope and design of the development, and on the location of public installations. The local plan must not, however, specify the intended purpose of premises so restrictedly that the premises lose a major part of their sale value. Thus a local plan cannot under protest from the owner allocate an area for, e.g. the production of cement, but it may do so for more general prupuses such as 'heavy industry' or the like.

Like the municipal plan, local plans are adopted by the local council after a public procedure which, however, is less comprehensive than that for municipal plans. Local plans which are in accordance with the municipal plan or possibly with a 'section-15 framework' may be implemented to begin with by the local council alone. However, the Minister of the Environment may, at his discretion, stop a draft local plan and decide that it shall have no legal effects until approved by him and, to safeguard national plan directives, the Minister can instruct the local council to provide a local plan with specified contents or provide a local plan himself.[9]

Pursuant to section 16 of the Municipal Planning Act, local plans are to be executed, first when necessary to secure the implementation of the municipal plan and second prior to carrying out large-scale developments or 'major building and construction works, including the demolition of buildings'. The latter passage, imposing a duty to provide local plans has had practical effects ever since the Act came into force on 1 February 1977 and many local plans have been worked out already. Local plans are comparable to some earlier types of plan, especially *town plans* which remain in force under the new Act until cancelled by a local plan.[10]

Directives in local and town plans can to some extent prevent residential areas from being troubled by pollution from industrial plants etc. but, as regards air pollution, these plans are relatively less important than for other types of pollution and they have little effect on the total air pollution. In this connection it may be mentioned, however, that in local and town plans areas may be zoned for recreative purposes (parks and the like) with the resulting faster ventilation and improved air quality in the urban area.

As shown above, national planning directives, regional plans, municipal plans and local plans form a joint planning system which is characterized mainly by the key words 'decentralization', 'overall control' and 'publicity'. Figure 2.1.(a) may serve as a rough illustration of the planning system.

Central level	National planning directives	The National and Regional Planning Act
Regional level	Regional plans	
Municipal level	Municipal plans	The Municipal Planning Act
'The citizens	Local plans	

Figure 2.1(a). The planning system.

Under the *Zone Act* the country is divided into *urban and rural zones*. (The Act adds a third, less important category, namely 'weekend cottage areas', but it may be disregarded in this context.) Urban zones are all areas which were zoned for urban development when the Act took effect on 1 January 1970, or were subsequently so designated. The final decision is usually taken in the form of a town plan or (after 1 February 1977) a local plan, but the directives of these plans concerning the zoning of an area for urban development will in future be comparatively closely governed by regional and municipal planning. All areas not zoned for urban development (or weekend cottage purposes) are rural zones.

The Zone Act contains no provisions as to which activities may be carried on in urban zones. Such restrictions on the owners' use will depend mainly on the contents of town and local plans. As regards rural zones, however, the Zone Act involves major controls. Pursuant to sections 6–8 of the Act, there is in principle a ban on development and new building and on buildings or areas being put to uses other than agriculture, forestry or fishing. This ban may—and is to a certain extent intended to be—set aside in specific cases where permission is granted by the county councils *under* guidelines laid down by the Minister of the Environment.[11] The county council is thereby enabled to influence the location of polluting plants in rural zones and areas surrounding heavily polluting plants may be kept clear. The forthcoming regional plans will have a binding effect on the county councils' administration of the said provisions. Control by local plans will be possible in rural zones too, although within narrower limits than for urban zones,[12] but in practice by far the most important control over permission to develop rural zones will be through the said provisions of the Zone Act.

As regards the *metropolitan area* (the municipalities of Copenhagen and Frederiksberg and the counties of Copenhagen, Frederiksborg and Roskilde), a separate act on regional planning has been passed.[13] It contains special provisions for regional planning in particular, but also other rules concerning ordinary physical planning unknown in the legislation for the rest of the country. Despite the differences, the basis of the systems in and outside the metropolitan area is the same, except that the power to prepare a regional plan for the metropolitan area is vested in the Greater Copenhagen Council whereas outside the metropolitan area it is vested in the county councils. Other differences between the two systems worth mentioning are that the regional plan for the metropolitan area must be somewhat more comprehensive than other regional plans, that the Greater Copenhagen Council can instruct a local council to complete a municipal plan by drawing up local plans and that this Council can forbid a local council to participate in private or public building developments within certain delimited areas.[14]

In connection with an account of the control opportunities afforded by the ordinary planning legislation, there is reason to mention *the system of building permits*. It is based on the *Building Act*[15] and, in its essentials, dates back to 1960 and even further back for towns. Under the Building Act no new buildings may be erected nor extension, rebuilding or demolition operations executed, or a change of use take place without prior permission being obtained from the local council in its capacity as building authority. When considering whether to grant a building permit, the local council shall first and foremost satisfy itself that the technical, constructional provisions of the Building Act and the building regulations have been observed and in that respect the system of permits is not relevant in this context. However, legislation increasingly provides that no building permit may be granted until the permits necessary under other legislation have been obtained or that the local council may at least postpone the grant of a building permit until directives administered by other authorities have been observed. Thus, before the local council grants a building permit, it must investigate whether the provisions of the Municipal Planning Act have been observed; whether a permit has been obtained from the county council for building in a rural zone; and, in cases coming under the Environmental Protection Act, whether the arrangement is in accordance with rules administered by the local council itself.[16] The system of building permits has thus acquired more general importance as a measure of control in relation to legislation concerning public-law control of real estate.

To sum up, it may be said that the general legislation concerning physical planning is of importance for the location of plants causing air and other pollution in three different respects. In the first place, local and town plans still in force provide legally binding frameworks for siting polluting plants. These frameworks are being increasingly determined by provisions at higher levels in the form of regional plans and national planning directives. Second, industrial plants which presuppose major building and construction operations may be established only when there is a local plan for the area. This means that a local plan procedure, which must involve the public, has to be set up first. Third, the planning legislation means that for the establishment of polluting plants in rural areas ('rural zone') permission must be obtained from the county council. In all current planning acts pollution is named as one of several aspects which are to be considered by the administrative authorities.

2.1.2 Specific environmental control as regards the location, arrangement and operation of plants

Whereas the above subsection dealt with controls aimed at environmental protection as only one of several objectives, the present subsection will concentrate on regulations specifically aimed at limiting and fighting pollution. Furthermore, the rules in this subsection concern not only the location of the plants, but also their arrangement and operation. The first subject discussed is the approval system provided for in the Environmental Protection Act, then the provisions of the same Act concerning the issue of specific orders and prohibitions and finally some general directives.

2.1.2.1 THE SYSTEM OF APPROVALS UNDER THE ENVIRONMENTAL PROTECTION ACT

As a central element in pollution control, Chapter 5 of the Environmental Protection Act introduced an approvals system. The tenor of this is that enterprises, plants and installations which may be described as 'heavily polluting' may not be established or commenced until approved. Similar rules apply to constructional and operational extensions or alterations which might increase the degree of pollution. The approval must take into account all types of pollution caused by the enterprise etc. in question. Decisions concerning the discharge of waste water must be made with reference to the special rules in Chapter 4 of the Act, but the decision to allow the discharge of waste water is taken at the time of the approval under Chapter 5 and the conditions on which the permit is granted are incorporated in the approval decision.[17] The approval system is important because it allows the authorities to make an overall evaluation of the environmental consequences of a polluting plant while it is still at the planning stage and to impose concrete requirements on it.[18]

An annexe to the Environmental Protection Act contained a *'schedule' of the types of plant, etc.* included in the approvals system. However, the Act authorized the Minister of the Environment to make alterations in the 'schedule' and so far this has been effected by a Government notice of 1978 containing the current regulations concerning the scope of the approval system.[19] The 1978 Notice does not differ significantly from the contents of the annexe to the Act, but is first and foremost a limited specification of the provisions in the original schedule based on practical experience.

The current 'schedule' comprises a total of 96 items, many of which cover several types of plant, etc. Compared with most foreign approval systems, the Danish system may be said to be far-reaching and to include a very large number of types of plant, etc. Thus the schedule comprises several types of enterprise which cannot in ordinary parlance be described as 'heavily polluting', for instance, fur farms, kennels, taxi ranks for at least 3 vehicles, and automatic car-washing plants. The reason is, of course, that when the schedule was drawn up not only the extent and dangerousness of the emissions were considered, but also, from experience, the nuisance caused thereby.

The schedule stresses only the nature and possibly also the size of the enterprise, plant or installation. It does not, however, distinguish between enterprises which are operated commercially and others which are not; and it makes no distinction between privately and publicly owned ones. The vast majority of plants etc. to be approved are of an industrial nature and the typical trade or agricultural undertaking is not included in the system. It should be noted, however, that 'poultry farms and pig farms are subject to approval'.[20] As regards installations usually under public ownership, there is reason to mention that the approvals system does not apply to purifying plants, roads and railways. Purifying plants are subject to county council control, however, pursuant to sections 21 and 22 of the Environmental Protection Act; under section 10 of the Act the Minister of the Environment can claim the right to give an opinion on road and railway projects before construction is commenced. For certain types of plant, etc. specified in the schedule approval has to be applied for only if the plant is not erected in accordance with a local or town plan. With the extended obligation to provide local plans following from the Municipal Planning Act, 1975, see 2.1.1 above, this modification in the compulsory approval of plants will have increased practical significance.

A considerable number of types of plant, etc. have been included in the schedule predominantly on account of the air pollution they may cause. By way of example, there may be mentioned cement works, power stations, fish-meal factories, disposal plants, slaughterhouses and fur farms.

The approvals system is directed first and foremost at the setting up of *new enterprises, plants and installations*. As mentioned above, the obligation to apply for approval also applies, however, to *constructional and operational* changes 'which involve increased pollution'. The fact that a plant changes to shift operation will thus definitely generate a duty to apply for approval unless previous approval of the plant has taken this into account.[21] On the other hand, enterprises, etc. 'in existence' when the Act came into force on 1 October 1974 need not apply for approval, see section 36 of the Act *per contra*. However, pursuant to this provision, the Minister of the Environment can lay down regulations to the effect that existing plants, etc. within one or more of the classes included in the schedule shall, within a certain specified time, apply for approval even though they are not being extended or changed. According to section 38 of the Act, the Minister can also direct specific plants to apply for approval with a certain time, 'if closer investigation of the pollution factors is deemed specially necessary' and this power exists whether the industry, plant or installation is included in the schedule *or not*.[22] Section 36 of the Act has not yet been invoked by the Minister of the Environment, however, and section 38 has so far only been applied once. Finally, it should be mentioned that it is possible to apply voluntarily for approval of existing plants, etc. covered by the schedule, see section 37 of the Act. It may be to the owner's advantage to have such approval because it provides significant protection against subsequent intervention by the environmental protection authorities.[23] In practice, however, only a few applications have been filed voluntarily for approval under section 37 of the Act and the majority have concerned cases where approval of an extension is being sought simultaneously.

In principle *decisions on approval* are made by the local council, see section 39, subsection 1 of the Act. However, pursuant to subsection 2, the Minister of the Environment may decide that this power shall be exercised by the county council or the Greater Copenhagen Council in regard to certain classes of plant, etc. In the above-mentioned schedule of enterprises, etc. covered by Chapter 5 of the Act, the Minister therefore provided that the power to decide on the approval of a number of types of plant, etc. should be delegated to the county council and, in the metropolitan area, to the Greater Copenhagen Council. The plants, etc.

in question are especially those which present a danger of regional pollution or pollution of a more complex nature (e.g. cement works and power stations) or which are subject to the authority of the county council and, in the metropolitan area, the Greater Copenhagen Council under legal provisions other than Chapter 5 of the Environmental Protection Act, e.g. Chapter 4 of that Act or the Raw Materials Act. In the case of a municipal installation, the decision as regards its approval is taken by the county council or the Greater Copenhagen Council, and in the case of a plant owned or operated by the county council or the Greater Copenhagen Council, the power of decision is vested in the Environment Agency.[24]

There is no indication in the Environmental Protection Act as to which considerations the authorities are to take as a basis when deciding whether to grant approval. This means that the authorities can use their discretion, i.e. after judging a concrete case they can choose the relevant aspects to be considered. This does not mean that they have a free hand, however. On the contrary, their *basis* of judgement is considerably restricted in many ways.

In the first place they are bound by the general provisions concerning the purpose of the Act, etc. contained in Chapter 1 of the Act. In this connection it should especially be emphasized that decisions on environmental protection pursuant to section 1 subsection 3 of the Act must be taken after weighing the nature of the surroundings and the effect of pollution on them against the benefit to the community of the enterprise and the cost of measures for environmental protection. Furthermore, it should be noted that anyone who wishes to set up a polluting industry according to section 3, subsection 1 of the Act, must choose such a place for the exercise 'that the danger of pollution is limited as much as possible' and that, pursuant to section 3 subsection 3 of the Act its operations shall be planned in such a way 'as to cause the least possible degree of pollution'.

Second, the Minister of the Environment can lay down general standards for emissions, etc. (sections 6 and 7 of the Act) and for the quality of surface waters, for the pollutant content of the air and for noise levels (section 8 of the Act). Standards laid down in pursuance of sections 6 and 7 of the Act can be made binding whereas standards concerning the quality of the environment under section 8 can only serve as guidelines. Binding regulations have only been issued in pursuance of sections 6 and 7 of the Act to a very limited extent so far.[25] On the other hand, in accordance with powers delegated by the Minister of the Environment, the Environment Agency has laid down a vast number of recommendatory standards concerning emissions, etc. As regards air pollution, such standards are to be found in *Vejledning fra Miljostyrelsen* (Guidelines from the Environment Agency), No. 7 1974 on the control of air pollution caused by plants, etc., No. 3 1976 on the control of air pollution caused by oil-fired plants and No. 2 1978 on the control of air pollution caused by plants emitting cellulose thinners and other compound thinners into the air.[26] These recommendations are particularly important in practice. Generally it can be said that the recommendations are followed except where concrete cases provide a basis for relaxing or for making the standards more stringent.[27] The reason for the practical impact of the recommendations is above all the expertise which lies behind the standards and the fact that the Environment Agency is the appeal authority for decisions by municipal authorities on environmental protection.

Third, the 'environmental quality planning' mentioned in 2.1.1 above might prejudice the basis of judgement in cases of approval. As already mentioned, however, no requirements as regards air purity have been laid down in this planning. Finally the decisions made by the Environment Agency and the Environmental Appeal Board in concrete appeal cases will in practice limit the latitude of local and regional authorities even further than do the general standards and environmental quality planning.[28]

The power of approval vested in the environmental authorities affords significant control over the most polluting plants, etc. *Approval* of a location applied for may be refused[29] and the approval may specify certain *conditions* as regards the establishment and operation of the enterprise.[30] In practice a number of conditions are usually attached to approvals.[31] Some of these conditions are very precise, e.g. a limit for the emission of sulphur dioxide; others are more flexibly formulated, e.g. a condition that there must be no serious nuisance to neighbouring property caused by smell.

In concrete cases where there are particular grounds for it, the approval may be granted for a limited period of time.[32] It is unlawful, however, to introduce a general time limit in approvals of specific types of enterprises, etc.[33]

2.1.2.2 SPECIFIC ORDERS AND PROHIBITIONS UNDER THE ENVIRONMENTAL PROTECTION ACT

As regards existing plants, etc. concrete orders for the remedy of pollution or injunctions against continued operation may be issued in pursuance of the Environmental Protection Act or of regulations issued under the latter. The regulations differ according to whether the enterprise, etc. is covered by Chapter 5 of the Environmental Protection Act and according to whether the plant is approved under this Chapter or not.

If it is an existing enterprise, plant or installation which is *included* in the *schedule* mentioned in 2.1.2.1 above, *but not approved* under Chapter 5 of the Act, the local council may, in pursuance of section 44, subsection 1, of the Act, issue orders for remedial measures to be taken if the enterprise involves 'pollution which is not deemed insignificant'. If the pollution cannot be remedied or if a prohibition is violated then, under section 44, subsection 2, the local council can issue an injunction against continued operation and, if need be, demand that the enterprise, plant or installation be removed. If the pollution causes 'imminent serious danger to health' an injunction under subsection 3 may be issued immediately. In the case of a plant, etc. operated by the local council or county council, or the Greater Copenhagen Council counting as the latter, competence is vested in the county council or the Environment Agency, respectively.

The basis for intervention by issuing orders—'pollution which is not deemed insignificant'—is very flexibly formulated. However, the discretion of the authorities is limited by the same types of factors as in cases concerning approval, see 2.1.2.1 above. There are two main modifications, however, which both tend to make the requirements in cases concerning orders and prohibitions less far-reaching than in cases concerning approval. In the first place, some of the standards in the guidelines apply to new plants only and where standards are laid down for existing plants they are usually less rigid.[34] Second, the benefit to the community of such plants or the costs involved in remedy will often weigh heavily in the case of orders and prohibitions, whereas in approval cases they will only be considered in quite exceptional circumstances.[35]

In cases of intervention against *plants, etc. approved under Chapter 5 of the Environmental Protection Act*, there is another and significant condition. Pursuant to section 44, subsections 1 and 2 of the Act can be issued in respect of such plants, etc. only if the pollution caused by the enterprise, plant or installation significantly exceeds that upon which the approval was based. In this connection it should be noted that in the approval the authority granting it must state the circumstances upon which the decision was based, see section 41 subsection 1. Section 44 subsection 4 thus affords the approved enterprise considerable legal protection against new environmental protection requirements.[36] This is largely reasonable, but it seems inexpedient to be unable, even after a number of years, to tighten up the environmental protection requirements for approved industries to take account of new technical knowledge concerning remedies and changes in opinion as regards the level of environmental protection.[37]

STATIONARY SOURCES OF AIR POLLUTION

On the subject of intervention against approved industries, it may be added that if the enterprise disregards a condition of approval then, in pursuance of section 42 of the Act, the approving authority may intervene by issuing an injunction against continued operation.

In the case of *an enterprise not covered by Chapter 5 of the Environmental Protection Act*, orders and prohibitions may be issued under the Environmental Protection Regulations laid down in pursuance of sections 4–5 of the Act.[38] Such an enterprise is popularly called a *'regulation enterprise'*. If such a commercially operated enterprise constitutes a serious nuisance to the environment, the local council can, under head 11 of the regulations, order that measures of remedy be taken. Where the nuisance cannot be remedied, the local council can issue an injunction against the enterprise. Plants wholly or partly owned or operated by public authorities are placed on the same footing as commercially operated enterprises. As regards nuisance caused by the keeping of animals and by furnaces, heads 2 and 10 contain similar regulations on orders and prohibitions and they do not distinguish between commercially operated enterprises and hobby activities. No immediate prohibition may be issued against a regulation enterprise owing to danger to health, as provided by section 44 subsection 3 of the Act, but then it is scarcely conceivable for a regulation enterprise to present an 'imminent, serious danger of health'.

2.1.2.3 GENERAL RULES

A number of general rules govern the arrangement and operation of plants solely with a view to preventing air pollution. The most important of these rules will be mentioned briefly. Furthermore, some general regulations will be dealt with which are aimed primarily at other objectives, but which have a considerable effect as regards air pollution.

In 1972 a special *Act* was passed, *limiting the sulphur content, etc. of fuels* which were used in stationary plants and in 1976 the scope of the Act was increased to comprise fuels used for purposes of transportation.[39] Essentially the Act does no more than authorise the Minister of the Environment to lay down limits for the content in fuels of sulphur and other substances which may cause air pollution through the use mentioned. The present rules concerning these limits appear in a government notice of 1976,[40] which provides that gas-oil (distillate oil), including diesel oil with a sulphur content exceeding 0·8% by weight, must not be used as fuel in stationary plants or for purposes of transportation on land or be sold for these purposes and that, after 1 October 1980, the percentage by weight must not exceed 0·5%. Fuel oil (residual oil) must not be used for firing in stationary plants or be used with a sulphur content exceeding 2·5% by weight. As regards the municipalities of Copenhagen and Frederiksberg and the county of Copenhagen, however, the sulphur content must not exceed 1% by weight. The Environment Agency may permit these maximum limits to be exceeded where, in concrete cases, it is deemed justifiable, owing to purification measures for example.

Pursuant to the *Building Act*[41] the Minister of Housing may lay down in national building regulations rules concerning the erection and arrangement of buildings having regard to 'fire-safety and health aspects'. In the building regulations—the latest being those of 15 January 1977—head 10 contains some requirements for the arrangement of furnaces and heating plants. The requirements are differential so that more stringent rules apply to large, closed furnaces and large central heating boilers (plants whose thermal output exceeds 60 kW) than to other furnaces and heating plants.

Finally, mention should be made of prototype approval of oil heaters. In pursuance of the *fire legislation*,[42] Notice No. 175 of 15 May 1963 provides that all oil heaters with fully or partly automatic control shall be manufactured in accordance with an approved system. To a certain extent similar rules apply to other oil-fired plants as well. Decisions on approval are taken by a testing committee for oil heaters set up by the Ministry of Justice, a committee whose decisions may be appealed against to that Ministry. The person or firm to whom prototype approval has been granted is responsible for ensuring that the heaters, etc. supplied are in exact agreement with the approval.

The regulations under the Building Act are not concerned solely with limiting air pollution; and the rules of the Fire Act aim only at limiting the danger of fire. Both sets of regulations have a considerable effect on the prevention of air pollution.

2.1.3 Requirements as to purifying processes, etc. prior to the discharge of pollutants into the air

2.1.3.1 ARRANGEMENTS FOR PURIFICATION

Section 3 subsection 3 of the Environmental Protection Act provides that anyone who carries on or wishes to embark on a polluting activity shall make the necessary arrangements for preventing and combating pollution and shall organize the operations so as to cause the least possible pollution. However, this is no more than a general statement of policy which is not legally enforceable.[43] There are at present no binding, general rules of law on the purification of discharges from sources of air pollution, but pursuant to section 6 of the Environmental Protection Act, the Minister of the Environment will be able to lay down directions on such purification. Requirements concerning purifying arrangements will frequently exist in individual cases, however. Such requirements may be imposed in conjunction with the approval of activities, etc. under Chapter 5 of the Environmental Protection Act or by orders issued in pursuance of section 44 of the Act or of the Environmental Protection Regulations (for further details, see 2.1.2 above). In this connection it should be mentioned that the authorities dealing with concrete cases of air pollution are not under any obligation to decide on the method of purification nor will they usually undertake to make such decisions. The authorities will normally concentrate their attention on the extent and nature of the emission and leave it to the person causing the pollution to choose the means whereby the pollution can be kept at an acceptable level. The polluter will usually be the person best placed to find the most suitable methods of purification for the lowest outlay and this arrangement will thus limit the amount of public as well as private resources spent on pollution control.[44]

Hence Guidelines No. 7/1974 from the Environment Agency on 'Limitation of air pollution caused by enterprises', which is dealt with in more detail in 2.1.4 below, is by and large confined to the statement of emission limits. On certain points, however, the Guidelines contain certain instructions regarding the arrangements for purification to be required. Thus with regard to kilns in cement works, for example, the guidelines say that 'dust separators should be designed with at least two independently operating units and be of such dimensions that the emission is never more than doubled if one unit is not working'.

2.1.3.2 CHIMNEY HEIGHT, ETC.

In accordance with the building legislation,[45] part 10 of the Building Regulations contains provisions on the dimensions and design of chimneys serving furnaces, etc. The rules are quite specific as regards small chimneys (chimneys serving furnaces, etc. in which the total thermal output does not exceed 120 kW and which under normal, continuous maximum load do not produce waste gas temperatures higher than 350 °C at the point of entry to the chimney) whereas the requirements as regards large chimneys are discretionary for individual cases.

Pursuant to section 6 of the Environmental Protection Act, the Minister of the Environment may lay down binding rules as regards chimney heights, among other things, but the legal powers have not been used to issue regulations in this field either. In two guidelines, however, the Environment Agency has established some recommended standards relating to chimney heights which supplement the general rules of the Building Regulations. The guidelines are addressed directly to the authorities to help them in their assessment of concrete cases and the guidelines have legal effect for individuals only in so far as is required in each case. This may be put into effect either by way of conditions for granting approval or by way of an order proper.

The recommended standards for chimney heights are contained in Guidelines No. 7/1974 from the Environment Agency on the limitation of air pollution from enterprises, section III, and Guidelines No. 3/1976 on the limitation of air pollution from oil-fired plants. Guidelines No. 7/1974 applies only to thirteen types of plant which are all covered by the system of approval under Chapter 5 of the Environmental Protection Act. In principle, the standards apply to new as well as to existing plants, but as regards the latter relaxations are often envisaged as reasonable. Guidelines No. 3/1976 covers oil-fired plants only, but then no distinction is made between plants scheduled under Chapter 5 of the Environmental Protection Act and other plants. To be more specific, this guideline concerns partly oil-fired plants fuelled by gas-oil (distillate oil) with a thermal output between 15 kW (13,000 kcal/h) and 30 MW (25·8 Gcal/h) and partly oil-fired plants with a thermal output of up to 30 MW for which fuel oil (residual oil) is used.

As regards gas-oil-fired plants, the standards concerning chimney heights are as follows:

Table 2.1(a) Chimney heights for gas-oil-fired plants

Thermal output, E	Height of chimney, H
I 15 kW < E ⩽ 120 kW (13,000 kcal/h < E ⩽ 100,000 kcal/h)	As stated in the Building Regulations for small chimneys
II 120 kW < E ⩽ 1 MW (100,000 kcal/h < E ⩽ 860 Mcal/h)	As stated in the Building Regulations for large chimneys. However, H shall be at least 1·25 × the difference in height between the chimney base and the roof ridge of any building within a distance of 100 m. In the case of flat roofs, however, the height of the chimney above the roof shall be equal to at least half the width of the building.
III 1 MW < E ⩽ 30 MW (860 Mcal/h < E ⩽ 25,800 Mcal/h)	As in II, but H shall, in addition, be equal to the difference in height between the chimney base and the roof ridge of any building within a distance of between 100 m and (100 + 15 × E (MW)) m of the chimney.

As regards fuel-oil-fired plants, the standards for calculating chimney height are considerably more complicated and will not be reproduced in this study. It should be added that fuel-oil-fired plants with a thermal output of more than 30 MW in industrial undertakings subject to approval, including power and district heating stations, are covered by Guidelines No. 7/1974, section 3.

2.1.4 Emission limits

Binding rules concerning limits for the composition, concentration and extent of emissions which may cause air pollution can be laid down by the Minister of the Environment under section 6 of the Environmental Protection Act, but no such regulations have been issued so far. In three Guidelines, viz. Nos. 7/1974, 3/1976 and 2/1978, the Environment Agency has set out some recommended standards for emissions from air-polluting plants. The purpose of these guidelines is to assist the authorities in their decisions about approval under Chapter 5 of the Environmental Protection Act and on orders under the same Chapter and under the Environmental Protection Regulations. In this connection see 2.1.2 above. To those who operate or intend to establish plants causing air pollution, of course, these standards also indirectly afford guidance as to the requirements they may normally expect to have to comply with.

Guidelines No. 7/1974 on the limitation of air pollution caused by industrial activities select 13 types of enterprise and plant from among the activities, etc. covered by Chapter 5 of the Environmental Protection Act. Examples of these are iron and steel works, cement works, factories producing asphalt and other materials for roads and some specified furnaces. For each of these types of plant, the guidelines state emission limits in the form of requirements concerning the total discharge of pollutants in a calendar month divided by the number of cubic metres of gas discharged coverted to Nm^3 (0 °C, 1 bar) of dry gas or the amount of product manufactured. In some cases, for example foundries, limits are set for new plants and also higher limits for existing plants, but otherwise the emission standards are intended, in principle, to apply to new and existing plants alike. It is, nevertheless, stated in the guidelines that even where particular limits are indicated for existing enterprises it will be reasonable in some cases to accept small deviations from the emission limits.

In connection with the emission limits, the guidelines contain some recommendations on measuring emissions from air-polluting plant and evaluating the results of the measurement. Such standards are important in practice as the measurement results usually have to be compared with limits laid down in an approval or contained in the guidelines.

Guidelines No. 3/1976 on the limitation of air pollution caused by oil-fired plants concerns oil-fired plants with a thermal output of up to 30 MW; for further information see 2.1.3.2 above. The guidelines state that no general limits are laid down as to how much pollution may be caused by an oil-fired furnace. The guidelines do, however, contain 'some standards which taken together give an idea of what may to-day be expected from a oil-fired installation which is well maintained and properly adjusted . . .' These limits include some standards as regards the soot content and the emission of oil coke including ashes. The standards differ according to whether gas-oil or fuel oil is used and whether the furnace is new or old (6–8 years). These guidelines also contain recommendations as regards methods of measurement.

Guidelines No. 2/1978 are much narrower in scope than the two former ones. They contain only standards relating to air pollution caused by industries which discharge cellulose thinners and other mixed thinners into the air. As regards new plants, the guidelines suggest that limitation of the emission should be effected if the total emission from the plants exceeds either 10 kg/h or 50 kg/shift.[46] If the plants exceed these limits, the concentration of the emission should not exceed 300 mg/norm. m^3. For existing plants limitation of the emission is desirable if the total emission from the plant exceeds either 15 kg/h or 75 kg/shift. If the emission from the plant exceeds these limits, the concentration of the emission should not exceed 300 mg/norm. m^3.[47]

Guidelines No. 2/1978 not only lay down recommended limits for emission, but also state the maximum acceptable *immission concentration contribution* which is understood to mean the contribution to the pollutants content of the environment caused by a given enterprise. In this respect the figure of 0·3 mg/m^3 is given as the limit, which is considered to correspond to the smallest concentration detectable by smell.

2.1.5 Duty of the polluter to monitor his emissions into the air

This question will only arise in the case of plants with a considerable pollution potential, i.e. in particular plants covered by Chapter 5 of the Environmental Protection Act. Such undertakings will often feel prompted to monitor and possibly measure their emissions continuously because otherwise they run the risk of being criminally liable for not complying with the demands made of them, see 2.1.4 above.

The duty to monitor and possibly measure with separate sanctions attached may exist in certain cases. Section 53 of the Environmental Protection Act imposes a duty on persons responsible for conditions or devices which may cause pollution to inform the local council if breakdowns or accidents cause substantial pollution or involve the danger thereof. This obligation could arise, for example, in the case of serious accidents in fertiliser factories involving the danger of poisonous gases escaping. Furthermore, it may be decided as a condition of approval under Chapter 5 of the Environmental Protection Act that the enterprise must carry out continuous measurement of the air pollution it causes and inform the local authority of the results.[49] In this connection the firm may be requested to keep a record of measurements stating certain specified information.[50]

2.1.6 Enforcement

2.1.6.1 MONITORING

It is primarily the local council which is the supervising authority as regards the observance of the *Environmental Protection Act* and the *Sulphur Content of Fuel Act* and decisions made in pursuance of these Acts. The council and persons authorised by the latter will have access to public and private property for the purpose of obtaining information. Furthermore, the supervising authority may require persons responsible for conditions which may lead to pollution to provide such information as is necessary to estimate and abate the pollution,[51] and a duty may be imposed on the person in question to supply samples of material and waste products, if any. In pursuance of section 56 of the Environmental Protection Act, the Minister of the Environment can lay down general rules concerning the control and supervision effected by local authorities. Such rules may, among other things, stipulate that certain undertakings are to be inspected at specified intervals and that particular methods of measurement are to be applied, in this connection also see section 9 of the Act which provides for prototype approval of measuring apparatus, etc.[52] No general rules concerning public authority supervision of air-polluting plant have been laid down so far, however.

Under the *fire legislation*, chimneys, etc. on buildings with a fireplace shall be inspected by an accredited chimney sweep at least once a year. The chimney sweep shall see that the chimneys do not cause unnecessary air pollution. If a chimney does, the environmental protection authorities must be notified.[53]

In 1978 the supervisory function of the chimney sweep was extended.

Because of the need to save energy the Minister of Housing was authorised, by an amendment to the Housing Act, to order obligatory control tests of the working order of heating and ventilating installations.[54] As from 15 December 1978, this authority has been invoked in a Notice.[55] The Notice provides that once a year for control purposes, the chimney sweep shall take measurements, for all small oil-fired installations, of waste gas temperature, carbon dioxide and soot. For the purpose of the Notice small oil-fired installations are atomiser-type oil heating installations with a thermal output not exceeding 60 kW for heating buildings and connected to chimneys covered by the requirement for regular sweeping.

The owner or user is notified in writing of the measured results and if the chimney loss exceeds 16% or if the soot content exceeds 3%, the chimney sweep will advise the owner or user that the installation requires adjustment and will also inform him/her of the economy in terms of oil resulting from such adjustment. If the owner or user has a contract with a service firm for periodical inspection he shall, at his request, be exempted from the chimney sweep's control.

The public authorities' surveillance is supplemented by private supervision on the part of persons or groups exposed to nuisance caused by air pollution. This kind of supervision is only of practical importance, however, in the case of air pollution by smell and dust.

2.1.6.2 SANCTIONS

The binding rules concerning the control of air pollution mentioned above are sanctioned by *penalties*. These apply to general directives in Acts and Notices including rules that permits shall be obtained, concrete orders or prohibitions and the observance of conditions on which permits are granted.[56] Criminal proceedings are instituted by the Director of Public Prosecutions usually at the instance of a body involved with the legislation on environmental protection. The case is dealt with by the ordinary courts in accordance with usual criminal procedure. The penalties imposed by the Environmental Protection Act are fines, simple detention or imprisonment for a period of up to 1 year and for offences committed by joint-stock companies or the like a fine may be imposed on the company as such.[57]

In the preparatory work for the Environmental Protection Act two different opinions on the sanctions provided in the Act are put forward.[58] On the one hand, it is argued that pollution cannot be controlled on the basis of sanctions and that educational and other motivating activities should be given higher priority. On the other hand, it is emphasised that there must be a possibility of applying sanctions, that it should be possible to impose penalties severe enough to have a real deterrent effect and that it is desirable to tighten up the former practice concerning penalties.

The first part of this objective has been fully achieved. Nobody can claim that needless punishment is meted out in pollution cases. In the case of minor, first offences against the environmental legislation, the only consequence normally is that the environmental authority reprimands the offender and may make the administratively established requirement more stringent. The reluctance to prosecute is clearly illustrated by the fact that during the period from October 1974 to December 1977 there were only some 150 criminal cases concerning violation of the Environmental Protection Act and regulations under it. The majority of these cases concerned unlawful discharge of effluent.[59]

The fact that there have been so few criminal cases is in itself an indication that it is doubtful whether the second part of the objective has been achieved. This assumption is confirmed by the fact that the level of the fines must be described as very low (rarely in excess of Dan. Kr. 1,000 or 2,000) and that there is no clear trend towards more severe treatment in cases where important environmental interests have been threatened. In this connection it may be mentioned that the level of penalties imposed in cases concerning environmental pollution is usually far below that in cases concerning tax evasion, defrauding the customs or contravention of the price legislation. In an August 1978 study, the Agency concluded that the penal usage could hardly be said to have been more severe since the coming into force of the Environmental Protection Act. At the request of the Ministry of the Environment, the Attorney General accordingly requested the prosecuting authorities in all types of environmental cases to try to make the usage concerning the level of fines stricter.[60] At the same time the Attorney General has stressed the need for cooperation between the Director of Public Prosecutions and the environmental protection authorities. In this connection it was particularly emphasised that the latter should assist the prosecution by providing particulars of the damaging effects of the offence and should be requested to comment on the sentence asked for by the prosecution.[61]

If, as a consequence of his punishable offence, the polluter has made a profit, this profit may be *confiscated*, possibly in conjunction with the sentence in pursuance of sections 75 and 76 of the Penal Code. The extent to which a saving made by the polluter through the offence may be confiscated in this way has not yet been established. So far, to the author's knowledge, there is only one criminal case—and that has not yet been finally adjudicated—in which, besides a fine, the prosecution has asked for the polluter's profit to be confiscated.

Even if penal practice is tightened up—and this seems to be under way—*administrative sanctions* will, in practice, be more important than penal ones. Such administrative sanctions may consist, in the first place, in the environmental protection authority *making an existing concrete decision more rigorous*. Violation of a permit may entail its withdrawal[62] and if an order is disregarded, it may afford grounds for the further operation of the plant being banned.[63] The deterrent effect of this, however, will depend on whether the Environmental Protection Authorities will in actual fact resort to such drastic measures. In a period of considerable unemployment, the authorities will probably be particularly reluctant to make decisions which will close down polluting enterprises—see the principle of considering the 'benefit to society' of the enterprise as provided in section 1 subsection 3 of the Environmental Protection Act.

If a decision on an order or prohibition has not been complied with within the period stated, the local council can also carry out the measure in question with respect to the party responsible and subsequently collect the amount by the usual legal means, see section 49 subsection 2 of the Act. In legal theory this coercive measure is referred to as '*self-help acts of the Administration*'. This means of enforcement which can, under the Environmental Protection Act—unlike most other acts—be applied without a preceding judgement, is very rarely resorted to, however.

2.1.6.3 SUSPENSION OF ENFORCEMENT

The party to whom orders and prohibitions are addressed may complain about the decision to at least one, and often two, superior administrative authorities. The lodging of such a complaint will under the Environmental Protection Act as a rule suspend the duty to observe the order. For further reference, see 1.7.4 above.

2.1.7 Air quality objectives

2.1.7.1 LEGAL REQUIREMENTS AND RECOMMENDATIONS

No binding quality requirements can be made as regards the pollutants content of the air. Section 8 of the Environmental Protection Act, however, authorises the Minister of the Environment to lay down recommendations concerning among other things the pollutants content of the air. No such recommended standards have been issued as yet, however.

In some respects the three Guidelines on air pollution mentioned above do contain some preconditions which may be described as quality objectives. It should be mentioned in particular that the recommendations as regards the calculation of chimney heights are based on the assumption that all immission concentration contributions from the plant in question as well as from other sources in the neighbourhood should not exceed 0·25 mg of particles/m^3 and 0·75 mg of sulphur dioxide/m^3 air measured as the mean for 30 minutes more than 15 times per month (corresponding to 1% of the time).[64]

2.1.7.2 DETAILS CONCERNING THE OBSERVANCE OF AIR QUALITY STANDARDS

A planned country-wide programme for measuring air quality has not yet been implemented, but is expected to be within about one year. For the time being, there are but few systematically collected and directly comparable results available from urban areas outside the metropolitan area. Since the mid-1970s a considerable number of measurements have been taken of sulphur dioxide and soot content in the metropolitan area. These measurements indicate that during the past decade the sulphur content of the air has fallen to approximately half and the soot content to about one-fifth of previous levels. Yet in both respects the quality objective set in the Guidelines has barely been achieved. However, since the beginning of the 1970s, the long-term aims of WHO on sulphur dioxide and soot levels have been observed. For further information on the development in air quality in Denmark, see the Environment Agency's publication *Miljøreformen*, 1979, pp. 109–19.

2.1.8 Rights of the individual

The individuals affected by decisions on environmental protection, including air pollution, may be divided into two groups: the polluters and the persons exposed to pollution. As regards the right of the former to protection under the law, readers are referred to 1.7 above. The same applies in the main to the persons who are exposed to pollution, but a few points regarding their position will be discussed below.

2.1.8.1 RIGHT TO INFORMATION

Under the Public Access to Documents Act, everybody has a right to inspect documents in administration cases if they can indicate the case in question. This means that, in principle, the public can demand to see material held by the authorities in a case concerning air pollution. This also applies to the results of air pollution measurements. Certainly, such documents will very often have been compiled by the staff of the authority in question and will thus be in the nature of internal material to which the public has no access but, from the rule of the above Act that verbal information which is important to the case must be made available to the public through the taking of notes, it may presumably be inferred that the same applies to important factual information which the authority has itself obtained by other means.

The right of the public to inspect documents is subject to exceptions, however. Of particular interest in this context is the fact that there is no access to information on the personal or financial circumstances of individuals nor to information on production or business circumstances if it is essential to the person or business concerned that such information is not disclosed to the public. As regards 'interested parties' in an environmental case, however, they will, in principle, have access to such information too, as it may be refused only if the refusal is based on 'decisive considerations of public or private interest'. As regards the concept of 'interested party' under the Public Access to Documents Act, see 1.7.2 above.

As a special rule on the access to information in the field of air pollution, mention should be made of the clause in section 4 of Notice No. 436 of 25 August 1976, according to which traders in oil containing sulphur are obliged, on demand, to give written information to consumers about the maximum sulphur content of the oil being supplied.

2.1.8.2 RIGHT TO APPEAL AGAINST PERMITS

Everybody who has an individual, essential interest in the outcome of the case has a right to appeal. This means that permits issued in respect of air-polluting plants may be appealed against by the neighbours of a particular individual plant. For further reference on this subject, see 1.7.3 above.

2.1.8.3 RIGHT TO MAKE STATEMENTS

Pursuant to Chapter 2 of the Public Access to Documents Act, 'interested parties' to an administration case have a general right to request that the decision be postponed until they have had the opportunity to make a statement. This right of interested parties to make statements involves an obligation for the authority, of course, to consider any such statement. For further reference, see 1.7.2 above.

2.1.8.4 RIGHT TO RAISE OR TO INTERVENE IN ENVIRONMENTAL CASES

Cases on environmental protection may be brought by the same people who are entitled to appeal, i.e. anybody who has an individual, essential interest in the outcome of the case. For further reference, see 1.7.1 above.

After a case concerning the issue of a permit has been raised, those entitled to appeal may, usually, under the Environmental Protection Act, suspend the legal effect of the permit by appealing against the decision to the Environment Agency. For further reference, see 1.7.4 above.

2.1.8.5 *LOCUS STANDI* (RIGHT TO TAKE LEGAL ACTION)

Individuals who are affected by a polluting plant are entitled to bring a legal action and may, by judgement, have an obligation imposed on the polluter to discontinue or at least limit his polluting activities or to pay compensation in respect of damage caused by pollution. For further reference on this subject, see 1.7.5–7 above.

2.2 MOTOR VEHICLES

Rules on air pollution caused by stationary plants are mentioned in 2.1. The remaing part of Chapter 2 deals with the standards for air pollution by mobile sources: 2.2 deals with the rules concerning air pollution caused by motor vehicles wholly or partly intended for use on the roads; and 2.3 deals with the problem of air pollution caused by other mobile sources—aeroplanes, ships, etc.

2.2.1 Design and equipment

Section 67 of the *Road Traffic Act*[65] contains a general provision to the effect that all motor vehicles shall be so designed and maintained in such order that they can be used without causing unnecessary danger or inconvenience to others (and without causing damage to the roads). More specific requirements with a view, among other things, to limiting air pollution by vehicles can be issued by the Minister of Justice under section 68 of the above Act.

These requirements are laid down, partly in a Notice on the design and equipment of motor vehicles,[66] and partly in accordance with section 44 of this Notice, in a Notice on detailed rules concerning motor vehicles.[67] Provisions concerning the requirements as regards air pollution caused by motor vehicles appear only in the detailed regulations which, under subsection 7.06, contain the following rules:

> For motorcars with *petrol engines* whose cylinder capacity exceeds 0·8 litres, the carbon monoxide content of the exhaust gas measured when idling must not exceed 4·5% by volume with an allowance of 1% by volume for any inaccuracy in measurement. For other motorcars, including *motorcars with diesel engines*, the smoke density in the exhaust gas must not exceed 3·5 units with an allowance of 0·3 units for any inaccuracy in measurement.

The detailed regulations mentioned apply only to vehicles which were registered, approved or taken into use for the first time in Denmark after 1 May 1977. By virtue of previous legislation, mainly the same rules apply to vehicles registered, etc. between 1 January 1971 and 1 May 1977. As regards older vehicles the requirements are somewhat less strict.[68] In addition to the specific regulations mentioned there is, as before, a general requirement that an engine shall be so designed as 'not to discharge unnecessary smoke'.

The Danish requirements mentioned concerning air pollution caused by motor vehicles are less stringent in several respects than the standards laid down by the EEC authorities.[69] The latter, in particular, cover discharges of carbon monoxide, hydrocarbons and oxides of nitrogen whereas the Danish rules concern only carbon monoxide and soot. In this connection, it should be noted that the EEC regulations require only that registration in member countries must not be refused to motorcars which comply with the EEC standards. This obligation has been complied with by Danish legislation.[70] The Environment Agency adopts the view, however, that the EEC standards should be introduced as standard requirements in Denmark as soon as possible.[71]

2.2.2 Maintenance

The requirements as regards the design of motor vehicles apply continuously as mentioned in 2.2.1 and the owner or person permanently in charge of the vehicle is responsible for its being maintained in such a way that it is always in regulation condition; in this connection, see section 67 of the Road Traffic Act. There are no special rules on *maintenance* to take account of air pollution.

2.2.3 Use

Air pollution caused by motor vehicles may conceivably be limited by various legal requirements concerning the use of motor vehicles, including restrictions as to the time and place of use. This type of regulation is represented by only one rule in Danish law, viz. paragraph 7 of part 12 of the Environmental Protection Regulations.[72] Pursuant to this rule, local authorities may, in order to prevent air pollution and nuisance from noise within specified areas, lay down rules restricting the right to allow an engine to idle while the motor vehicle is stationary. There are no known examples, however, of this rule having ever been applied by local authorities and its practical use is very limited, to say the least.

2.2.4 Requirements concerning the fuel used in motor vehicles

By an Act of 1977, the Minister of the Environment was authorised to lay down rules limiting the content of lead compounds and other substances in petrol used in motor vehicles and introducing a system of supervision and control over importers and producers of and dealers in petrol.[73] In pursuance of this, the Minister issued a Notice containing regulations effective from 1 January 1978 concerning *the content of lead compounds in petrol*.[74] According to these provisions, petrol with a content of lead compounds above 0·40 g/litre in terms of lead must not be imported or produced for the purpose of being used as fuel for motor vehicles. The Ministry of the Environment may waive this rule, however, if it is deemed necessary in order to secure reasonable supplies of petrol for the country, or if other special grounds dictate it.

Prior to the implementation of these provisions the lead content was not limited by law but, for some years prior to the 1977 Act, an 'agreement' existed between the oil companies that the lead content must not exceed 0·84 g/litre of petrol. It is estimated by the Environment Agency that the reduction of the permitted lead content in petrol as from 1 January 1978 helped to reduce the total annual emission of lead in Denmark from some 900 tonnes to some 700 tonnes and that bringing the lead content limit down to 0·15 g/litre may reduce the emission to barely 300 tonnes per year.[75]

2.2.5 Enforcement

Observance of the provisions of the *Road Traffic Act* relative to preventing air pollution is ensured first and foremost by requirements relating to inspection and approval before a motor vehicle is registered for the first time. For most vehicles this is effected by a prototype approval.[76] The requirements regarding air pollution must be complied with at all times and not only at the first registration. This may be ensured by the police calling in a particular vehicle for inspection according to section 77 and 78 of the Road Traffic Act, or, as regards vans or lorries and vehicles for commercial conveyance of passengers, self-drive hire or ambulance services, etc. through periodic inspections.[77]

It should be emphasised that there is at present no legal authority for calling in all motor vehicles for periodic inspections or for doing this on the basis of general criteria. As regards the majority of motor vehicles, subsequent inspection can be demanded only if specific circumstances afford grounds for the police to require that a vehicle be inspected. Further, it should be noted that at subsequent inspections of motorcars the inspectors will normally only examine the emission of air pollutants in the case of vehicle models which often present problems in that respect. The supervision is exercised partly by the State Motor Vehicle Inspectorate in connection with the inspection and approval of motor vehicles, and partly by the police as part of their general supervision of the observance of road traffic legislation.

The supervision as regards the observance of the rules pursuant to the *Lead Content of Petrol Act* is much simpler because the legal requirements are aimed at the importers and producers. Pursuant to the 1977 Notice, importers or producers of petrol for use in Denmark are under an obligation to give the Environment Agency information in writing on demand concerning the lead content in specific stocks or deliveries, and the Environment Agency, or others authorised by the latter, may take samples of stocks of petrol held by importers and producers.

Violation of the regulations laid down in pursuance of the Road Traffic Act is *punishable* by fines. The provisions of the Lead Content in Petrol Act are also enforceable by law, and violation of them may be punished by fines or imprisonment. If an importer or producer makes an incorrect written declaration to the Environment Agency on the lead content in petrol, the penalty may under section 162 of the Penal Code be up to 4 months' imprisonment.

Criminal cases are dealt with by the ordinary courts according to general rules of criminal procedure. As far as the writer is informed, there have been no criminal cases in this field as yet.

2.2.6 Rights of individuals

As far as the rules mentioned in 2.2 above are concerned, only the polluter is, in general, considered an interested party; nobody else has an individual and essential interest in the observance and application of the rules. In other words, no 'third party' can have the rights of an interested party in this field. This state of affairs has various consequences.

Thus, a 'third party' will *only be entitled to information* according to part 1 of the Public Access to Documents Act which means that all the exemption rules of that Act apply. In this connection, see 1.6.3 above for further details. Nobody other than the person responsible for compliance with the legal requirements—the owner, importer or manufacturer of the vehicle—has any *right to appeal against decisions made by administrative authorities* (in this connection, see 1.7.3 above) or any *right to make statements* before a decision is made; for further reference, see 1.7.2 above. Nor can 'third parties' *bring actions* whether to obtain the overruling of an administrative decision, the prohibition of polluting activity or damages; for further reference, see 1.7.5–7 above. In this connection it should especially be noted that violation of the rules concerning the design and equipment of motor vehicles and the lead content in petrol cannot in practice form the basis of an action for damages and that the rules of the general law of adjoining properties do not apply to this sector as they only concern nuisance issuing from real property.

The one who is responsible for observance of the rules—the owner, importer or manufacturer of the vehicle—has, on the other hand, the ordinary rights of an interested party, in this connection see 1.7.1–7 above. As regards the right to appeal, it may be appropriate to mention that the owner or user of the vehicle can appeal to the Ministry of Justice against decisions made by the State Motor Vehicle Inspectorate or the police, and that decisions by the Environment Agency in pursuance of the Lead Content of Petrol Act cannot be appealed against to any other administrative authority.[78]

2.3 AEROPLANES, HOVERCRAFT AND SHIPS

2.3.1 Aeroplanes

Under the Air Traffic Act, all aircraft must be registered.[79] As a condition of registration, the Minister of Public Works can, under section 9 of the

AIR

above Act as amended by Act No. 94 of 29 March 1972, impose requirements for the purpose of abating nuisance caused by noise or otherwise affecting persons outside the aircraft. These powers have not yet been used to lay down regulations concerning air pollution caused by aeroplanes, however. Section 6 of the Environmental Protection Act can, if necessary, also be used to define limits to discharges from aircraft as well as regulations concerning aircraft design, operation and maintenance.

Within the meaning of the Environmental Protection Act, airports and aerodromes are in the nature of 'specially polluting enterprises' and, as such, they are subject to approval by the county council in accordance with Chapter 5 of the Act; in this connection, see 2.1.2 above. On the requirements which may be imposed as regards the siting of airports and use of surrounding land, refer to 8.1.[80]

2.3.2 Hovercraft

There are no special provisions regarding air pollution or other pollution caused by hovercraft, which are used to only a small extent in Denmark. The Minister of the Environment could, however, issue regulations of this kind in pursuance of section 6 of the Environmental Protection Act which concerns emission limits, design, operation, maintenance and prototype approval.

In this connection, mention should be made of a general provision as regards *ships* in section II of the Standard Harbour Regulations which reads 'No craft may, while moored or sailing in harbours discharge so much smoke as to cause a nuisance to those living in the neighbourhood or to the traffic in or near the harbour'.[81] This regulation extends to hovercraft. Supervision of compliance with the regulations is undertaken by the local harbour authority which can issue defining orders. Harbour regulations are enforceable by law, but criminal proceedings relating to section II of the Standard Harbour Regulations have, as far as the writer is aware, not been instituted so far. On enforcement and the rights of individuals, refer to 2.2.6.

2.3.3 Ships

Readers should refer to the account under 2.3.2.

Notes

1. Statutory Notice No. 513 of 29 September 1978.
2. Statutory Notice No. 494 of 19 September 1975, as subsequently amended.
3. Act No. 287 of 26 June 1975, as subsequently amended.
4. Circular of 28 February 1978 from the Ministry of the Environment.
5. For further details, see *Nyt fra planstyrelsen* No. 5/1978.
6a. As amended by Act No. 288 of 26 June 1975.
6b. Circular of 22 February 1977.
7. For further details, see Claus Tonnesen, in *Dansk Miljøret*, vol. 3, 1977, p. 45 ff. and the circular mentioned in 6b.
8. See the circular of 17 April 1974 on waste water and the circular mentioned in 6b, part 3.1.2.
9. See sections 26, 28 and 29 of the Municipal Planning Act.
10. For further information on the Danish planning act system, readers are referred especially to Vagn Rud Nielsen and Bent Andersen, in *Dansk Miljøret*, vol. 2, 1977, pp. 7–246.
11. In this connection, see circular of 25 April 1973 from the Ministry of Housing.
12. See section 18, subsection 3 of the Municipal Planning Act.
13. Statutory Notice No. 128 of 16 March 1978.
14. For further details, see Bent Andersen, in *Dansk Miljøret*, vol. 2, 1977, pp. 44–56.
15. Statutory Notice No. 530 of 25 October 1976, as subsequently amended.
16. See section 16, subsection 3 of the Building Act, and, for further details, the Building Regulations of 15 January 1977, part 1.3 with enclosure 2.
17. See Notice No. 176 of 29 March 1974, section 7. On the relation between Chapters 4 and 5 of the Act, see also Chapter 3 of this book.
18. What follows gives only a summary description of the approval system. Concerning the detailed contents refer especially to the Notice mentioned in note 17; Circular of 15 May 1974 from the Ministry of the Environment; and to Claus Haagen Jensen, *Godkendelse af forurenende anlaeg*, 1977, pp. 18–62.
19. Notice No. 290 of 28 June 1978. To a very limited extent this Notice was amended by Notice No. 154 of 6 April 1979 on dust-extraction plant for manually controlled smoothing machines in car painting shops.
20. On the definition of 'pig farms' compared with ordinary farming, refer especially to the decision by the Environmental Appeal Board, in *KFE* 1976, p. 7.
21. See Circular of 15 May 1974 from the Ministry of the Environment, part 5, item 2.
22. For further details on the interpretation of this provision, see Claus Haagen Jensen, *Dansk Miljøret*, vol. 3, 1977, p. 59 ff.
22a. Cf. the Environment Appeals Board decision of 6 October 1980, *KFE* 1981, p. 151.
23. See section 44, subsection 4 of the Environmental Protection Act, and 2.1.2.2 below.
24. See section 39, subsection 3 of the Act and section 6 of Notice No. 176 of 29 March 1974.
25. Probably the only existing relevant example as regards air pollution is Notice No. 154 of 6 April 1979 on dust-extraction plant for manually controlled smoothing machines in car painting shops.
26. For further details, see 2.1.3.6 below.
27. For further details see Claus Haagen Jensen, *Dansk Miljøret*, vol. 3, p. 69 ff.
28. For further details, see *op.cit.*, pp. 65–76.
29. For example, see letter of 8 January 1976 from the Environmental Appeal Board, *KFE* 1976, p. 147, and letter of 19 December 1978, *KFE* 1979, p. 78.
30. For further reference, see Claus Haagen Jensen, *op.cit.* pp. 77–80 on the right to impose different conditions, including generally formulated nuisance conditions.
31. By way of example, refer to letter of 25 May 1978 from the Environmental Appeal Board, *KFE* 1979, p. 18 and *Nyt fra Miljøstyrelsen*, No. 7/1978.

32. By way of example, see letter of 17 October 1978 from the Environmental Appeal Board, *KFE* 1979, p. 60.
33. See *Nyt fra Miljostyrelsen*, No. 5/1976, p. 22.
34. As an illustration, see the Environment Agency's Guidelines No. 7/1974 on the limitation of air pollution by industrial plants, section II.
35. In this connection, compare the letter from the Environmental Appeal Board of 19 November 1976, *KFE* 1977, p. 58, with letter from the same authority of 10 October 1975, *KFE* 1976, p. 145.
36. For further details on the interpretation of section 44, subsection 4 of the Act, see Claus Haagen Jensen, *op.cit.*, pp. 92–96.
37. See Claus Haagen Jensen, *op.cit.*, p. 97 ff., where it is contended that it ought to be possible to revise conditions of approval after a period of 10 years, for example, as is the case in Swedish law on the subject.
38. Notice No. 170 of 29 March 1974, as subsequently amended.
39. Act No. 135 of 26 April 1972, as amended by Act No. 259 of 26 May 1976.
40. Notice No. 436 of 25 August 1976; for further details see Circular of 25 October 1979 from the Environment Agency.
41. Section 6 of Statutory Notice No. 530 of 25 October 1976.
42. Statutory Notice No. 641 of 29 December 1976.
43. In this connection, see Section 83 of the Environmental Protection Act concerning penalties.
44. For further details of the modest requirements in regard to determination of both conditions and permits and orders in environmental cases, see Claus Haagen Jensen, *op.cit.*, pp. 78–80 and 86–89.
45. See 2.1.2.3 above.
46. By shift, the Guidelines mean 8 hours.
47. When calculating the emission limits the Environment Agency took into account, *inter alia*, the relevant German standards (*Technische Anleitung zur Reinhaltung der Luft* of 1974) but nevertheless felt that the standards laid down by it should be somewhat more lenient.
48. The Environment Agency took the German standards (VDI 2306 of 1966) as a basis, but in this respect the Danish standards are stricter than the German ones.
49. See Notice No. 176 of 29 March 1974, section 10.
50. See the Environment Agency's Guidelines No. 7/1974, section III, subsection 7.
51. For further details of the polluter's duty to provide information and the limits thereof, see Claus Haagen Jensen, *op.cit.*, pp. 114–21.
52. Recommended standards for this purpose are to be found, for instance, in the Environment Agency's Guidelines No. 7/1974, section IV.
53. See Instructions No. 124 of 29 May 1970.
54. Act No. 610 of 14 December 1977.
55. Notice No. 608 of 4 December 1978.
56. As regards the Environmental Protection Act, see section 83 therein.
57. The Act concerning the limitation of the sulphur content, etc. in fuel stipulates fines or detention as the penalties and according to a number of notices, only fines may be imposed. In the case of very serious violations of environmental legislation, it may also be possible to punish for contravention of the provisions of Chapters 20 and 21 of the Penal Code concerning various acts that endanger or harm the general public, for which the penalty may be up to 10 years' imprisonment.
58. See *Folketingstidende* 1972–73, appendix A, col. 4020 ff.
59. See The Environment Agency's book, *Miljøreformen*, p. 57, November, 1979.
60. Memorandum No. 5/1979 of 9 October 1979 from the Attorney General.
61. According to the publication mentioned in note 59, p. 57 ff., a certain tightening up of the policy of the courts in cases of violation of the Environmental Protection Act, particularly in cases concerning effluent, was observable even prior to the issue of the Attorney General's memorandum. On the basis of a study of the 150 or so

criminal cases mentioned above, Mr. Sven Ziegler, judge, concluded that in his opinion there was no reason generally to criticise the fining policy, *Politiken*, 28 May 1980.
62. See section 42 of the Environmental Protection Act concerning approvals under Chapter 5.
63. See 2.1.2.2 above.
64. See Guidelines No. 7/1974, section III, 6, and Guidelines No. 3/1976, section III, 3.5.
65. Act No. 287 of 10 June 1976, with subsequent amendments.
66. Notice No. 154 of 20 April 1977 with subsequent amendments.
67. Latest Detailed Regulations on Vehicles, 1980. These regulations are not published in *Lovtidende* where Acts and the majority of the Notices of regulations are otherwise published.
68. In this connection see 2.2.1 of the first edition of this account.
69. See Directive 70/220 of the Council of Ministers as amended by Directive 72/306, Directive 74/290 and Regulation 77/102 of the Commission.
70. See Notice No. 451 of 1 August 1973.
71. See p. 209 of *Miljøreformen*, November 1979, published by the Environment Agency.
72. Notice No. 170 of 29 March 1974.
73. Act No. 267 of 8 June 1977.
74. Notice No. 354 of 21 June 1977.
75. *Nyt fra miljøstyrelsen*, No. 1/1980, p. 18 ff.
76. For further reference, see Notice No. 161 of 26 April 1977, part 2.
77. See section 78, subsection 1 of the Road Traffic Act and part 9 of the Notice mentioned in note 72, and circular No. 142 of 28 June 1966.
78. The former follows from the general, unwritten rules of administrative law on appeal and the latter from section 2, subsection 2 of Notice No. 354 of 21 June 1977, together with section 4 of Act No. 267 of 8 June 1977.
79. Statutory Notice No. 381 of 10 June 1969, with subsequent amendments.
80. On air pollution caused during the years 1968–70 by aircraft at Kastrup airport, which is the only international airport of any importance in Denmark, refer to 2.2.2 of the first edition of this account (1976).
81. The legal basis of the Harbour Regulations is now section 19 of Act No. 329 of 12 May 1976, on commercial harbours according to which the Minister of Public Works lays down working rules for each harbour by negotiation with the harbour authority in question. No regulations have been laid down as yet under that Act and until then section 20, subsection 6 of it upholds the existing regulations.

3
Fresh Water

The environmental reform of 1973 brought about a collation of the legal regulations regarding the protection of fresh water against pollution. Before this the regulations were to be found mainly in the Water Supply Act, the Watercourse Act and the local health by-laws and were administered in part by different authorities, but after 1 October 1974 the regulations were concentrated mainly in the Environmental Protection Act and were administered under it by the general authorities. Nevertheless, the previous legislation left its mark on the rules in the Environmental Protection Act; thus the Act's most important provisions concerning water pollution in Chapters 3 and 4 are based on a distinction between the protection of water supply interests and the protection of surface water. This distinction, which does not seem appropriate in every respect, does not fit readily into the plan of this study. It should also be noted that there are still a number of regulations outside the Environmental Protection Act that are relevant in the present context, see in particular 3.2.

Chapter 4 of the Environmental Protection Act on the protection of surface water relates not merely to the restriction of discharges into watercourses and lakes but also to discharges into the sea. As the regulations are similar in all essentials, a number of questions concerning marine pollution are dealt with in this section, insofar as they come under the provisions of the Environmental Protection Act.

3.1 STATIONARY SOURCES OF WATER POLLUTION

3.1.1 Control through general planning legislation

Readers are referred to 2.1.1, which gives a brief account of the legal instruments contained in the general planning legislation and the connection between this planning and the special 'environmental quality planning' under sections 61 and 62 of the Environmental Protection Act.

It is important to note that, whereas no real environmental quality planning has been implemented in respect of air, since 1975 provisional targets have existed for water quality in surface waters within the individual counties.[1] In pursuance of the Water Supply Act, moreover, a significant amount of work has been done under the direction of the county councils on charting the position, size and quality of water resources and planning the future utilisation of water.[2] This has involved the direct designation of areas important for the supply of water in the future and, indirectly, of areas that are of no interest in this respect. The fact that in pursuance of section 59, subsections 1 and 3, of the Water Supply Act quality requirements have been laid down concerning surface water which is to be used to produce drinking water[3] and for drinking water and water that is used in conjunction with food production,[4] respectively will have an important impact on future water supply planning. Both these sets of quality requirements correspond to standards adopted by the EEC.[5] Finally it must be added, while discussing the significance of the planning legislation in relation to environmental protection, that probably the most important effect on water pollution is that the planning legislation ensures that future building development is concentrated thus making sewerage and the purification of waste-water a practical possibility.

3.1.2 Preventive control

The most important controls on water pollution relate to the direct discharge of substances which may pollute the water, whether groundwater or surface water (lakes, watercourses and sea) is involved. This control is referred to in greater detail below, under 3.1.4 in particular.

One's first thought is that the controls should be directed solely against the direct discharge of polluting substances into water. But in this as in many other spheres, effective protection cannot be achieved exclusively by means of a direct defence system and it is therefore appropriate for this to be extended and supplemented by preventive regulations. Hence there are in Danish law a number of important regulations aimed at preventing the discharge of polluting substances into water. This is done through requirements being imposed on plants, depots and activities which are known from experience to present a danger of water pollution owing to their nature or situation. The majority of these requirements are designed to protect water supply interests but the object of some regulations is to prevent the pollution of surface water which is not intended for the water supply.

3.1.2.1 PROTECTION AREAS IN CONNECTION WITH WATER SUPPLY INSTALLATIONS

The extraction of water normally involves a permit. The decision on this is taken by the local council if it is a question of small-scale extraction of groundwater—generally not more than 3000 m^3 per annum—and for other extraction projects by the county council.[6]

When the county council gives permission for the extraction of groundwater, it may under section 12 of the Environmental Protection Act lay down a protection area within which cesspools, etc. receiving WC effluent, and possibly wells, boreholes and subterranean containers are to be prohibited upon expiry of a specified time limit. A similar protection area may be established under section 13 of the Act near municipal watercourses or a lake with the effect that no commercial undertakings, institutions, camping sites or the like must be operated in the area and there must be no deposits of substances which may pollute the water supply installation. The county council can, moreover, under section 14 of the Act, issue orders and prohibitions to prevent the risk of pollution of water supply installations which have been granted permission by the county council under the Water Supply Act. With regard to the extraction of ground water, the county council is also empowered to issue orders and prohibitions with a view to protecting future water supply installations. To illustrate requirements that may be imposed under section 14, orders can be cited to relay waste-water pipes, ban the use of certain fertilisers on neighbouring land and ban sailing on watercourses.[7]

3.1.2.2 SUBTERRANEAN CONTAINERS FOR OIL, ETC.

Section 11, subsection 2, para. 2, of the Environmental Protection Act contains a ban on the unauthorised installation underground of containers holding liquids and substances which may pollute the groundwater except for watertight containers used solely for liquid manure or silage effluent.[8] This requirement applies to both new and existing containers. The Minister of the Environment is also able, under section 11, subsection 2, to lay down general rules to ensure that the groundwater is not exposed to pollution by liquids and substances, including regulations concerning the control and installation and emptying of containers holding the said liquids.

In pursuance of these statutory provisions more detailed regulations have been laid down concerning the storage of oil, etc. and collecting tanks. The regulations covering the storage of oil are now in Notice No. 171 of 29 March 1974 and relate chiefly to subterranean *oil tanks*. In respect of oil tanks already installed or that might be installed underground after 1 April 1970 (when the first environmental protection regulations concerning oil tanks came into force), the Notice sets out a number of requirements concerning, in particular, the construction of the tanks and their distance from water supply installations. The most important provisions in the Notice are those to the effect that the tanks shall be of a type that has been approved by a central testing committee and that the building authority (normally the municipal authority) shall be notified thereof at least 8 days before the tank is covered in. The notification must be accompanied by a certificate from the seller stating, among other things, details of the tank and that the tank is of an approved type. The requirements concerning the construction, etc. of the tanks vary according to whether the tank has a capacity of up to 6000 litres, between 6000 and 100,000 litres or over 100,00 litres. Guidelines on the construction of types of oil tank with capacities up to 100,000 litres form an annexe to the Notice.

For tanks installed underground before 1 April 1970, the general rule is that the user must ensure, at the latest 20 years after installation, that the tank fulfils the requirements for tanks installed underground after 1 April 1970 or he must take special protective measures (effective cathode protection, etc.) and make subsequent regular checks. With regard to tanks holding more than 100,000 litres, however, the Environment Agency establishes special conditions in each individual case both for new and old tanks.

Should an oil tank be found to be leaking, the municipal authority can require it to be emptied and dug up and any contents that have leaked out and polluted the soil surrounding the tank must be removed. Orders of the latter kind may also be served on the oil supplier if substantial amounts of oil have escaped during filling.

In accordance with section 16, subsection 1, of the Notice, the municipal authority may, with the Environment Agency's approval in special cases where the water supply is not affected, allow the requirements of the Notice to be modified. Under this provision the obligation to replace oil tanks installed underground before 1 April 1970, at the latest 20 years after their installation, was partly waived for a period from the spring of 1978 to the spring of 1980 because of the heating planning in progress.[9]

Notice No. 173 of 29 March 1974 contains rules about *collecting tanks for domestic effluent* installed underground. By 'collecting tanks' the Notice means watertight containers for the collection of domestic effluent. The municipal authority may permit the use of collecting tanks for the discharge of domestic effluent from houses containing one or two households. Various conditions are imposed for granting such permission; among other things, the medical officer of health has to recommend that permission be granted, various constructional provisions must be complied with and the tank must be located at a certain distance from water supply installations, buildings and roads. The Notice further requires that the owners must participate in any joint municipal disposal scheme and, in other instances, prove to the municipal authorities that the tank is being emptied and that the contents are properly disposed of. Permits of this kind may, in pursuance of section 11, subsection 3, of the Environmental Protection Act be altered or withdrawn at any time in the same way as other permits issued under section 11, subsection 1, if sewerage makes another form of discharge possible or if other environmental protection considerations make it desirable.

3.1.2.3 OTHER PREVENTIVE CONTROLS

Section 17 of the Environmental Protection Act, which is dealt with more fully below under 3.1.4, contains *inter alia* a ban on substances that may pollute water being stored so near watercourses, lakes or the sea that there is a risk of their being washed into these. There can be no dispensations from this ban.

A number of general preventive clauses are also to be found in the Environmental Protection Regulations.[10] These relate, among other things, to refuse, the construction of animal houses, dungheaps, fertiliser tanks and silage containers. The Minister of the Environment is able to issue further general preventive regulations in pursuance of various statutes. Under section 11, subsection 2, of the Environmental Protection Act, the Minister can lay down regulations to ensure that groundwater is not polluted by liquids and substances. The Minister is entitled to draw up preventive regulations with a broader scope under sections 6 and 7 of the same Act. Of particular relevance is section 7 concerning the right to prohibit or restrict the import or use of specific substances

STATIONARY SOURCES OF WATER POLLUTION

and this rule has been used to issue several preventive Notices, which are dealt with below in section 10 covering products.[11]

In Guidelines No. 1/1975 the Environment Agency laid down some recommendations about restricting pollution from *aquaculture*. Under point III of this there are guidelines for the operation and maintenance of fish farms, including those on, for example, the use of fish foods. The provisions are only recommendations but might with regard to compensation be of legal significance as standards of what might be deemed reasonable conduct.

Finally, it should be mentioned that requirements for preventing water pollution may be imposed as specific conditions for obtaining permission to discharge waste-water under section 18 of the Environmental Protection Act or approval under Chapter 5 of the Act for specially polluting enterprises. In this context it is pointed out, in particular, that most installations for the storage, deposit or processing of *refuse* have to be approved under Chapter 5 of the Act, about which there is more in section 7 below.

3.1.3 Pretreatment

The only general requirement existing in relation to the treatment of waste-water before it is discharged into the ground, surface waters or sewers by which the water is conveyed to surface waters, concerns discharges into soakaway installations, see 3.1.4.1.2 below. In accordance with Notice No. 139 of 15 April 1980 relating to the discharge of waste-water into the ground, domestic effluent must pass through a sedimentation tank before it is discharged into a soakaway installation.[12] Furthermore, in individual cases rules concerning pretreatment will often be established as conditions of discharge permits, so for instance requirements are laid down concerning the disinfection of waste-water from hospitals, laboratories and the like before it is discharged into the sewerage system.

Concerning the use of purification plants in conjunction with the collective discharge of sewage, see section 6.4.

3.1.4 Control over the composition and amounts, etc. of effluent discharged

The central element of defence against water pollution consists of two far-reaching bans. First, substances that may pollute water must not be discharged into surface waters (section 17 of the Environmental Protection Act) and, second, polluting liquids must not be discharged on or into the ground (section 11 of the Act). With these bans go permit schemes which are discussed below under 3.1.4.1. In respect of existing discharge and conveyance arrangements, there are also powers to intervene in order to prevent or limit the pollution, see 3.1.4.2 below. In 3.1.4.3 the criteria for granting permission and for subsequent intervention will be detailed. Finally, some aspects concerning charges in respect of the discharge of waste-water, etc. will be touched on under 3.1.4.4.

3.1.4.1 PERMIT SCHEMES

3.1.4.1.1 Conveyance to surface water

As mentioned, section 17 of the Act provides, in general, that substances which may pollute water must not be discharged into watercourses, lakes or the sea (or be stored so close to them as to create a risk of their being washed into the latter). The scope of the ban is extremely wide. Thus it covers the discharge of any substances which may pollute surface water (i.e. have harmful ecological effects on it), whether the substances involved are solid, liquid or gaseous and all forms of discharge, irrespective of whether this occurs through actual discharge or via rainwater from, for instance, roads.[13]

It goes without saying that the ban is not absolute. It is envisaged that, by and large, permits will be granted for the discharge of waste-water and, moreover, permission may exceptionally be granted by the Environment Agency for, say, the use of chemical agents in surface waters for combating weeds or for scientific purposes.

With regard to *waste-water discharges* into surface water the arrangement is that any discharge—direct or indirect—requires prior permission from a public authority. The specific rules concerning the *distribution of competence* appear in the *Sewage Notice*[14] issued in pursuance of the Environmental Protection Act, which is supplemented by a *sewage circular*, also drawn up by the Ministry of the Environment.[15]

Under the Sewage Notice, the municipal councils always grant permission for connection to or alteration of discharge to public sewage installations within the predetermined capacity of the latter, connection to or alteration of the discharge of domestic effluent to private sewage plants within the predetermined capacity of the plant and the installation of new or the extension of existing private sewage-plants for domestic effluent, insofar as the system is not designed to receive waste-water corresponding to more than 30 person equivalents (approximately 10 households). When a sewage plan has been approved for the municipality (see below), the municipal council's powers are extended to include permits for the establishment or extension of public sewage installations or major private installations for domestic effluent within the framework of the plan and also to permit the discharge of waste-water other than domestic effluent from enterprises, etc. (including roads) which are not covered by Chapter 5 of the Environmental Protection Act. In all other cases the county council decides on permission to discharge effluent into watercourses, lakes or the sea. In particular, the county council always grants permission for the special discharge of waste-water into watercourses, lakes and the sea from enterprises operated by a municipal council.[16]

Individual conditions may be attached to permission for waste-water discharge, etc.[17] In special cases which are specified, the permission may be made subject to a time limit.

The rules for permission mentioned are in many respects more stringent than those valid before 1 October 1974 and they are, under section 19 of the Act, applicable only to waste-water installations that have been legally executed under the previous legislation to the extent and from the date determined by the Minister of the Environment. No firm decision on this has yet been taken.

3.1.4.1.2 Discharge into the ground

As mentioned, section 11, subsection 1, of the Environmental Protection Act contains a ban on surface water, drainwater, waste-water and other liquids which may pollute the groundwater being discharged above ground or discharged underground via cesspools, etc. without permission. Permits are issued under the Act by the Minister of the Environment but he may, in accordance with sections 45 and 46 delegate his powers, as has happened through the *soakaway notice*.[18]

If it is a matter of the *discharge of domestic effluent to soakaway installations* (cesspools, etc.), the municipal council may, for properties outside built-up areas, permit discharge from at most two households or of an amount of effluent corresponding to 10 person equivalents at most. What is meant by a property being situated 'outside a built-up area' is that within a distance of 300 m of the dwellings, etc. on the property there are similar buildings for no more than four other properties. Permission may be granted, however, only if a number of more specific conditions are fulfilled, see section 4, subsection 1, and when permission is given some listed requirements must be fulfilled or their observance must be aimed at, see section 4, subsections 2 and 3. The conditions relate chiefly to the distance between the soakaway installation and water supply installations.[19] For properties in built-up areas permits are issued for the same restricted discharge arrangements as those taken by the county council as its basis, but the latter may subdelegate its powers to the municipal authority in accordance with specified guidelines. The discharge of domestic effluent from more than 10 person equivalents may be permitted within certain limits by the county council, whose powers in this respect cannot be further delegated. With regard to the *discharge of waste-water other than domestic effluent to soakaway installations*, the competence to issue permits is divided between the municipal council and the county council. The municipal authority may, on certain conditions, allow the discharge of roof-water, water from paved areas (apart from public roads, car parks and paved areas in conjunction with enterprises under Chapter 5)[20] and waste-water from milking parlours. The county council may give permission for the discharge of waste-water from gravel washing and from waterworks' filters and surface water from public roads, car parks and paved areas in connection with enterprises under Chapter 5. The discharge of waste-water above ground, for instance by spraying or by irrigation, may be permitted by the county council if the discharge involves effluent that is biodegradable and contains only substances and nutrient salts that may easily be metabolised.[21]

Outside the framework laid down in the Notice for permits to be issued by the municipal authorities and county councils, permission to discharge waste-water may be granted only by the Environment Agency. In cases of this type the Environment Agency acts as the primary authority, whereas otherwise the Agency only deals with cases concerning discharge into the ground as the appeal authority.

3.1.4.1.3 Situation with regard to Chapter 5 of the Environmental Protection Act

If a 'scheduled enterprise' is involved, i.e. an enterprise, etc. which has to be approved under Chapter 5 of the Act, the decisions about discharges are nevertheless taken in accordance with Chapters 3 and 4 and regulations issued in pursuance thereof. The Notice on approval,[22] however, ensures that competence is normally vested in the same authority and that the two decisions are technically linked—concerning approval of the enterprise, etc. and discharges of waste-water, etc.—since the conditions for the latter have to be incorporated in the approval under Chapter 5.[23]

3.1.4.2 INTERVENTION IN EXISTING DISCHARGE ARRANGEMENTS, ETC.

If polluting substances are being fed into surface waters, an order may be issued for the improvement or replacement of an existing waste-water installation if it is not operating properly from an environmental viewpoint, see section 25 of the Environmental Protection Act. This provision may be invoked whether permission had previously been granted for the installation under Chapter 4 of the Environmental Protection Act or under previous statutory regulations, or whether it had been lawfully established without permission. The decision concerning the order is taken by the authority, county council or municipal council which is competent under section 18 and the Sewage Notice to reach a decision on issuing a waste-water permit. In addition, it will presumably be possible to withdraw permits issued under Chapter 4 of the Act in accordance with general unwritten rules of administrative law if the conditions are seriously or repeatedly violated.

Permits for discharge via cesspools, etc. and into the ground may, under section 11, subsection 3, of the Environmental Protection Act be altered or withdrawn at any time if sewerage renders another method of discharge possible or environmental protection considerations make it desirable. This rule applies to both permits under section 11 of the Act and those issued under previous legislation.

Neither section 25 nor section 11, subsection 3, imposes special restrictions on intervention in respect of installations permitted under Chapter 4 or 3, respectively, of the Act. On the other hand, this is done by section 44, subsection 4, of the Act, under which orders and prohibitions concerning enterprises, etc. approved under Chapter 5 may be issued only if the pollution is substantially in excess of that taken as a basis for the approval.[24] In the legal debate after the Environmental Protection Act was passed, it was discussed whether the Act should be interpreted in such a way as to exclude the possibility of intervention concerning waste-water discharge in the case of approved scheduled enterprises, except in accordance with the provisions in section 44, subsection 4, of the Act. The predominant view in the theoretical opinions on the issue was that such an interpretation of the Environmental Protection Act was not tenable and that section 25 must take precedence over section 44, subsection 4.[25] In support of this viewpoint it may be argued that it would be unreasonable to afford scheduled enterprises better protection against requirements concerning changes in waste-water discharge than others that have obtained permission for the discharge of waste-water. It would, in particular, be invidious if section 44, subsection 4, could prevent a change in the requirements imposed on a scheduled enterprise which discharges waste-water into a common installation, and as such a restriction might well impose a greater purification burden on others discharging effluent into the installation. As yet no definitive attitude has been adopted to the question in administrative practice.[26] It should be added that with regard to the discharge of waste-water by scheduled enterprises as permitted under Chapter 3 of the Act, no-one has maintained that section 44, subsection 4, ought to restrict the issuing of orders as required.[27]

3.1.4.3 MATERIAL CRITERIA FOR DECISIONS CONCERNING DISCHARGE

Specific decisions on the discharge of waste-water, etc. are regulated substantially by three different factors: environmental quality planning, the sewage plans and recommended general standards.

As mentioned previously under 3.1.1, planning with regard to recipient quality has been executed in the individual counties. This planning may still be described as provisional and is taking place with reference *inter alia* to the sewage plans and the quality standards laid down under the EEC. Nevertheless, environmental quality planning provides the starting point for the specific decisions on discharge, since these have to be based primarily on the effect of the discharge in relation to the requirements for the general condition of the recipient concerned.

Under section 21 of the Environmental Protection Act, an overall plan has to be executed for each municipality for the development of local waste-water installations ('the sewage plan'). A draft sewage plan is prepared by the municipal council for approval. In pursuance of the Sewage Notice the municipalities should have drawn up draft sewage plans before 1 October 1976 and sewage plans have now actually been approved in more than 150 municipalities. For further details about the content and development of sewage plans, see below under 6.2.

There are as yet no binding general regulations in this sector but a number of recommended standards are contained in the Environmental Agency's Guidelines No. 6/1974 (sewage guidelines) and also to some extent in the Sewage Circular.[28] These rules are discussed more fully below, a distinction being made as to whether the recipient is water (watercourses, lakes or sea) or the ground. Questions concerning discharge into a waste-water system are dealt with in principle in Chapter 6.

In the case of domestic effluent, by which is meant waste-water from households, including water closets, no more specific requirements are imposed with regard to what may be discharged. It is a prerequisite for other permits that the individual discharge permit should indicate requirements concerning the composition, amount and temperature of the discharge. Therefore the rules mentioned below are relevant only in the latter case.

3.1.4.3.1 Protection of water recipients

The rules in the sewage guidelines may be divided into two categories: requirements concerning recipient quality and requirements concerning the individual discharges. Among the first should be mentioned the recommendatory provisions based on recipient quality and recipient limits (*immission limits*), see subsections 2.4.5.–15 of the Guidelines. These are drafted in a fairly abstract way. As an example, the provision in subsection 2.4.5 is to the effect that the discharge should not contribute substantially to the eutrophication of the recipient and in subsection 2.4.13 that the oxygen content in the surface layers of watercourses, lakes and sea should at no time be less than 5 mg/l.

More precise rules are stated for the individual discharges. In this context the guidelines provide a table of recommended limits for discharge to recipients (*emission limits*). The aim of the requirements is to achieve water quality corresponding to pollution level II (beta-mesosaprop), i.e. a water quality allowing salmon to live and breed, when certain other conditions are fulfilled. The emission limits are as follows, a dash in the schedule indicates that the Environment Agency has been unable to indicate a general level:

Table 3.1.4.3

Parameter	*a* Lakes and watercourses to lakes and enclosed fjords	*b* Watercourses to open bays, sounds, straits and sea	*c* Enclosed fjords and other enclosed salt and brackish waters	*d* Open bays, sounds, straits and sea	
pH	6·5–8·5	6·5–8·5	6–9	—	(1)
Temperature	30°C	30°C	30°C	35°C	(2)
BI$_5$	20 mg/l modified	20 mg/l modified	100 mg/l	400 mg/l	(3)
COD (Chemical oxygen demand)	—	—	—	—	(4)
NH_3-N: ($NH_3^+NH_4^+$)	2 mg/l	—	—	—	(5)
Total N	—	—	—	—	(6)
Total P	1 mg/l	—	1 mg/l	—	(7)
Sediment	0·5 ml/l	0·5 ml/l	1 ml/l	1 ml/l	(8)
Floating objects		Should not occur to any visible extent			
Suspended matter (total)	30 mg/l	30 mg/l	80 mg/l	—	(9)
Mercury	—	—	—	—	(10)
Cadmium	—	—	—	—	(10)

FRESH WATER

Table 3.1.4.3—continued

Parameter	a Lakes and watercourses to lakes and enclosed fjords	b Watercourses to open bays, sounds, straits and sea	c Enclosed fjords and other enclosed salt and brackish waters	d Open bays, sounds, straits and sea	
Chromium ($Cr^{III}+Cr^{VI}$)	0·2 mg/l	0·2 mg/l	0·2 mg/l	0·2 mg/l	
Copper	0·1 mg/l	0·1 mg/l	0·2 mg/l	0·5 mg/l	
Zinc	0·5 mg/l	0·5 mg/l	1 mg/l	1 mg/l	
Lead	0·1 mg/l	0·1 mg/l	0·5 mg/l	0·5 mg/l	(11)
Nickel	0·2 mg/l	0·2 mg/l	0·5 mg/l	0·5 mg/l	
Silver	0·05 mg/l	0·05 mg/l	0·05 mg/l	0·1 mg/l	
Arsenic	0·5 mg/l	0·5 mg/l	0·5 mg/l	1 mg/l	
Cyanide	0·1 mg/l	0·1 mg/l	0·1 mg/l	0·2 mg/l	(12)
Hydrogen sulphide	2 mg/l	2 mg/l	5 mg/l	—	
Free chlorine	0·3 mg/l	0·3 mg/l	0·5 mg/l	—	(13)
Halogenated phenols	—	—	—	—	(10)
Phenols	0·2 mg/l	0·2 mg/l	0·2 mg/l	—	(14)
Stable oil emulsions of mineral oil	5 mg/l	5 mg/l	5 mg/l	10 mg/l	(15)
Anionic detergents (80% degradable)	2 mg/l	2 mg/l	5 mg/l	10 mg/l	
Other synthetic detergents	—	—	—	—	(16)
Halogenated hydrocarbons	—	—	—	—	(17)
Organic solvents	—	—	—	—	(18)

Looking at the schedule in general, it reveals that the requirements presuppose that the concentrations indicated are reduced to one-tenth within the discharge zone fixed (dilution zone) and that it should not be permissible for the requirements to be met through the addition of neutral water from, for instance, waterworks. The following comments in note form apply to the individual points in the schedule:

(1) *Re d*: the levels were fixed taking into account the buffer capacity of the waste-water and the dispersion involved in the recipient. The discharge should not cause any substantial change in pH outside the dilution zone.

(2) The discharge of large amounts of cooling water should not nor-

mally cause the temperature of the recipient outside the permitted discharge zone to differ fundamentally from that obtaining naturally.

(3) *Re d*: if non-toxic and hygienically acceptable waste-water is involved, higher limits may be accepted in certain cases.

It may be remarked, more generally, concerning BI_5 that the oxygen content of the surface layers of the recipient must at no time be less than 5 mg/1 and that the biochemical oxygen consumption measured as BI_5 in fresh and brackish water should not exceed 6 mg/l (unmodified) immediately outside the discharge zone.

(4) Limits should be fixed for COD where it is a question of anything other than domestic effluent. After a period of parallel measurements of BI_5 and COD, testing may be carried out as a COD check only.

COD is determined by means of potassium dichromate.

(5) In the winter period it may be necessary to tolerate minor excesses.

(6) The discharge should not contribute substantially to the eutrophication of the recipient.

(7) Higher phosphorus levels than those indicated should only be tolerated on the basis of a recipient investigation. See too, note 6.

(8) Measured after being allowed to stand for 2 hours. The levels indicated here are applied regardless of the degree of dilution. It should be pointed out, moreover, that the discharge should not give rise to permanent deposits of sediment around the place of discharge or elsewhere in the recipient.

(9) See note (8) on sediment deposits at the place of discharge.

(10) Specially noxious substances which should not, in principle, be allowed to be discharged into surface water.

(11) Any discharge of metals should be limited as much as possible.

Discharge permits should contain quantity restrictions for metals (g/24 hours, kg/year).

If discharges from major communal systems are involved, it may to some extent be necessary to accept higher concentrations and amounts of zinc than stated for the present.

(12) *Re d*: the level is fixed according to the current state of the recipient with a maximum of 2 mg/l.

(13) The amount of free chlorine discharged into fresh waters should

be reduced as much as possible, even when the waste-water is disinfected with free chlorine.

(14) *Re d*: the level is fixed according to the current state of the recipient, with a maximum of 1 mg/l.

(15) There should be no visible traces of oil.

(16) The discharge permit is governed by biodegradability and toxicity.

(17) All discharges of halogenated hydrocarbons should be limited as much as possible. The group includes a number of *specially noxious substances*, see note 10.

(18) The biodegradability and toxicity of the substances varies a great deal and should be determined by biological testing. If it is found that *specially noxious substances* are involved, refer to note 10.

3.1.4.3.2 Protection of soil recipients

The Sewage Circular in 4, II, E contains some rules on this subject that are formulated in an abstract way. Among other things, the county council when granting permission for the discharge of waste-water should require that the quantity discharged into the ground annually does not substantially exceed 3500 m^3 per hectare (10,000 m^2). See too 3.1.4.1.2 above concerning the county council's powers in respect of such disposal.

The EEC directive concerning protection of groundwater against pollution caused by certain dangerous substances, which was adopted by the council on 17 December 1979, will not necessitate changes in Danish legislation.

3.1.4.4 CHARGES FOR WASTE-WATER DISPOSAL

The users of common waste-water installations must pay contributions towards their establishment and operation. However, these contributions should, in principle, correspond to the actual costs of executing and operating the installation, see expressly section 27, subsection 2, of the Environmental Protection Act, as amended by Act No. 107 of 29 March 1978. There is no 'charge' on waste-water disposal for *purposes of environmental control* under current law. The introduction of an actual charging control system in the environmental protection sector has been considered, in particular by a committee which was set up by the Minister of the Environment after the Environmental Protection Act had been passed by the Folketing. The committee's report dated August 1975 shows that the committee was unable to recommend the use of charges as an instrument of environmental policy, either to replace or supplement the use of regulations.[29]

In regard to contributions towards joint waste-water installations, see 6.2 below for further details.

3.1.5 Control of dumping in fresh water or on nearby land

Under section 17 of the Environmental Protection Act, substances which may pollute water must not be deposited in watercourses, lakes or the sea or stored so close to these that there is a risk of them being washed into the water. In regard to the possibility of waiving this ban by permits, readers are referred to the account above under 3.1.4. It should be added, moreover, that Danish watercourses and lakes are so limited in size, that dumping in them is not a practical possibility.

3.1.6 Supervision on the part of the polluter

The person who is responsible for circumstances or installations which may cause pollution must, under section 53 of the Environmental Protection Act inform the municipal authority if operating disturbances or accidents cause substantial pollution or involve the danger thereof. This obligation comes into play *inter alia* if a significant amount of pollutant from a plant escapes into a watercourse.

In addition, certain general supervisory regulations apply to underground oil tanks.[30]

Furthermore, supervision on the polluter's part may and in many cases will be incorporated in the conditions of the individual waste disposal permit. These may include requirements concerning how and how often the disposer has to take samples, their analysis and the submission of the analysis results to the supervisory authority.

3.1.7 Enforcement

The Environmental Protection Act contains some wide-ranging general provisions concerning *supervision* by public authorities. These are to the effect that municipal authorities must generally see that the Environmental Protection Act and regulations issued under it are complied with and that the county council should supervise installations operated by the municipal authorities and also the state of pollution of watercourses, lakes and the coastal sections of the territorial waters.

More detailed and more precise supervisory regulations may be laid down by the Minister of the Environment under section 56 of the Act and such provisions were issued concerning the supervision of waste-water installations and the pollution of areas of water by Notice No. 177 of 29 March 1974.[31]

In accordance with this, the municipal council supervises private waste-water installations and the observance of regulations designed to protect the water supply. The county council has the job of supervising public waste-water installations and the state of pollution in watercourses, lakes and in the coastal sections of the territorial waters. In the metropolitan area the county council's supervisory functions are performed by the Greater Copenhagen Council. Waste-water installations that are operated or maintained by the county council or the Greater Copenhagen Council come under the supervision of the Environment Agency. In order to safeguard special fishery interests, supervision of fishery controls is also carried out under the Ministry of Fisheries. If violations of the environmental protection legislation are found during such supervision, the ordinary supervisory authority is informed accordingly. The above-mentioned Notice also sets out certain provisions concerning the nature and scope of the supervision. Thus it is laid down that for public waste-water installations, major private communal waste-water installations for domestic effluent and waste-water installations with separate discharge from industrial plants, the recipient shall normally be inspected twice yearly around the outflow.

The above regulations concerning water pollution are all subject to *penal sanctions*. As regards the application of penalties and of other measures against unlawful polluting conduct, readers are referred generally to the account under 2.1.6.2 above. It is specially pointed out that by far the majority of all actions brought under the environmental protection legislation have concerned the unlawful discharge of pollutants into surface water. Within this group of cases approximately 75% of the sentences passed and fines imposed related to discharges from agriculture, chiefly in the form of silage effluent and liquid manure.[32] The latter circumstance helps to account for, though not necessarily to justify, the relatively low level of fines hitherto.

3.1.8 Water quality targets

3.1.8.1 LEGAL REQUIREMENTS AND RECOMMENDED GUIDELINES

A feature of the Danish system is that the targets for the quality of water recipients are laid down for the individual watercourses and lakes in the county's environmental quality planning. In this connection see 3.1.1 above. Provisional recipient quality plans of this nature have existed since 1975. The targets are fixed by the individual county councils and are not completely comparable. In general the aim for most watercourses appears to be that it should be possible to use the watercourse for fishing. For a number of watercourses, however, the target is simply that the watercourse should be aesthetically satisfactory.

General nationwide quality standards exist only to a very limited extent. *Binding* quality standards can, within the scope of the Environmental Protection Act at present, only be established for water used for bathing, see section 4 of the Act compared with section 8. On the other hand, the Water Supply Act 1978 contains in section 59, subsection 1, the authority to lay down binding regulations concerning the quality of groundwater and surface water which is to be used for water supply purposes. In pursuance of this the Minister of the Environment laid down in Notice No. 162 of 29 April 1980 regulations concerning quality requirements for surface water which is used to produce drinking water. The regulations it contains accord with the EEC's council directive 75/440. In this context it should also be mentioned that in pursuance of the Water Supply Act, section 59, subsection 3, Notice No. 6 of 4 January 1980 laid down binding provisions concerning the quality of drinking water and water that is used in connection with food production, corresponding to the standards in the EEC directive which was approved in principle by the council on 19 December 1978. Both these sets of regulations are very important for water supply and environmental quality planning but they do not embody direct general requirements for the quality of water recipients. Nor do they contain *recommended* quality standards to any great extent. It should be mentioned that the Environment Agency's Guidelines No. 6/1974 concerning waste-water contain some vaguely worded general preconditions relating to the quality of water recipients, see 3.1.4.3.1 above. The requirements of the guidelines may, in general, be said to express the desirability of a recipient quality corresponding to 'good fishing water'.[33]

The reluctance to lay down general standards of quality is due to two circumstances in particular. First, the issue of quality standards requires prior groundwork in natural sciences which has only been carried out in recent years or is still in progress. Second, the environmental policy of Danish legislation since the environmental reform in 1973–74 has been for environmental requirements to be laid down on the basis of the quality of the individual recipient and the target for this.[34] This line of thought is reflected, *inter alia*, in section 1, subsection 3, of the Environmental Protection Act and in the fact that under section 8 of the Act only recommended quality standards may be laid down.

It is envisaged, however, that in the coming years many quality standards will be issued relating to water, among other things, both in the form of binding regulations and recommended guidelines. In any case, the evolution of EEC participation in the environmental sector will bring about a substantial change in or modification of the tactics adopted in Danish environmental policy hitherto. Thus the implementation of the EEC's fishing waters directive of 18 July 1978, 78/659, will necessitate before July 1985 an amendment of the law which will enable quality standards to be established. It is further pointed out that the Environment Agency intends to issue in the course of 1980–81 new guidelines concerning waste-water, which are to supplement Guidelines No. 6/1974 and which are to a higher degree based on general recipient quality standards.

3.1.8.2 INFORMATION ON OBSERVANCE OF THE WATER QUALITY TARGETS

Since general quality standards do not yet exist by and large, there is no question of observance so far as the regional recipient quality plans are concerned, and there is insufficient data to be able to assess on a nationwide basis whether the targets for the individual watercourses and lakes have been realised.

As regards the trend in the quality of fresh water, readers may refer to the Environment Agency's report *Miljøreformen* of November 1979, pp. 135–44 and 194–202, which shows, among other things, that since the environmental reform there has been a not insubstantial improvement in the state of pollution, especially in a number of the most polluted watercourses.

3.1.8.3 RESTRICTIONS ON THE USE OF WATER WHICH DOES NOT MEET THE QUALITY REQUIREMENTS

As far as can be seen, the only relevant provisions are those concerning quality requirements for drinking water and water used in connection with food production, contained in the Ministry of the Environment's Notice No. 6 of 4 January 1980, see 3.1.8.1 above.

3.1.9 Individuals' rights

In regard to the individual's right to information, to be heard, to intervene in enforcement proceedings and to bring an action, refer to the account given above under 2.1.8 and also to Chapter 1. It is only the individual's right of appeal and the possibility of obtaining damages that call for special comments.

The decisions of the local authorities and county councils concerning the discharge of polluting substances and liquids into the ground and to surface waters may be *appealed* against to the Environment Agency which has, as a general rule, the final administrative say in the matter. Further appeals may be lodged with the Environmental Appeal Board only in respect of decisions on the discharge of waste-water into surface waters (sections 18 and 25 of the Act) if these related to the waste-water installations of a municipality or a commercial undertaking, decisions on the discharge of waste-water into the ground by commercial undertakings (section 11, subsections 1 and 3) and decisions on orders and prohibitions preventing the danger of water supply installations being polluted (section 14, subsection 1).[35] In regard to the extent of those entitled to appeal, refer to section 1.7.3.

The question of *damages* for injury caused by water pollution is often of practical relevance and by far the majority of actions for compensation in respect of pollution damage have related to losses suffered through water pollution or the risk thereof. Should losses of this kind occur, the injured parties may be able to obtain damages from the polluter. In the absence of legal regulations providing otherwise, the basis of liability follows the general rule on *culpa*.[36] However, in legal practice there has been a clear tendency towards making the liability stricter.

In the first place the assessment of *culpa* has become quite stringent. Considerable demands are made on the property-owner's expertise, his equipment and supervision of potentially polluting installations and activities. If there are preventive regulations in the sector, the violation of these will practically always imply grounds of liability. Judgement is particularly harsh if damage caused by dangerous substances is involved.[37] Perhaps the legal position re the basis of liability may be described by saying that, should an injury caused by water pollution (in the wide sense dealt with here) be ascertained, the property-owner or the person otherwise responsible for the installation is liable in damages, unless the injury may be deemed completely unforeseeable.[38] The basis of liability is not objective at present but the state of the law comes very close to objective liability.[39] Second, a tendency may be detected in some cases to relax the requirements for documentation of a causal association between the water pollution ascertained and an injury that has occurred.[40]

The most important recent judgements concerning liability in damages for water pollution injuries are U 1958 365 H, U 1963 806 H, U 1969 923 ØL, U 1971 672 VL and U 1977 183 VL.[41] In all these legal decisions the polluter was found liable in damages. Apart from the relatively few printed judgements, there are a large number of agricultural tribunal rulings concerning liability in damages for water pollution injuries.[42]

It may be worth taking a closer look at U 1977 183 VL because this concerned a practical situation that had not previously occurred in a printed judgement. Owing to an accident involving the electric pump on a factory's oil tank, oil flowed out and sank into the ground. The accident was discovered on a Sunday by a neighbour, at whose insistence the municipality immediately took steps to combat the pollution that had occurred on both the factory's land and adjacent property. During the legal action the municipality claimed costs for preventing and combating the oil pollution, to be payable by the factory owner. The municipality's claim was allowed in respect of the efforts outside the factory property on the basis of the general *culpa* rule and in respect of what was done on factory property under the rules concerning business management.

In a 1978 article on liability in damages for environmental damage caused by buried chemical waste, Bernhard Gomard raised the question *inter alia* of whether the environmental authorities can claim reimbursement from the landowner of costs involved in tracing and analysing buried waste.[43] Gomard's main conclusion was that the authorities could demand reimbursement of these costs if buried waste on the property presented a pollution hazard for groundwater. This view is based on the authorities being able, under section 11, subsections 1 and 3, of the Environmental Protection Act, to require the buried waste to be removed whether its burial was originally legal or not. The assumption may be subscribed to as far as unlawfully buried waste is concerned, see above on U 1977 183, but not otherwise. However, from the above-mentioned provisions in section 11 of the Environmental Protection Act it can not be concluded that the general unwritten rule may be broken that public authorities bear the costs of their own investigations unless this follows from special statutory provisions concerning a duty to inform.[44]

Certain decisions reached in order to protect against or to limit water pollution will affect lawfully established activities in such a way that the intervention is tantamount to expropriation. This means that the injured party has an undoubted claim to damages, even though there is no question whatever of negligent conduct. Intervention of this nature may occur in conjunction with decisions under sections 13 and 14 of the Environmental Protection Act concerning the protection of water supply interests when the owner in question is deprived of the hitherto legal exploitation of his property.[45]

3.2 POLLUTION OF FRESH WATER BY SHIPS

It should be noted initially that sailing on fresh water in Denmark is of negligible practical significance as a source of water pollution. The watercourses are, almost without exception, too small to take anything other than canoes, small rowing boats and the like. On a few lakes there is a certain amount of yachting but this does not give rise to any major water pollution problem.

3.2.1 Shipbuilding and fitting out

Requirements relating to the building and equipment of ships, etc. appear in the Shipping Supervision Act[46] and in the Building and Equipment Notice issued in pursuance thereof.[47] These provisions apply to cargo vessels that are rated at 5 tons and over and passenger vessels irrespective of tonnage but pleasure vessels of under 20 tons are not normally covered by the regulations. This means that the rules concerning ship building and fitting out are of little practical relevance to the vessels that navigate lakes and watercourses.

3.2.2 Maintenance

The requirements mentioned under 3.2.1 continue to apply and there are no special provisions as to maintenance.

3.2.3 Use

The main provisions governing this are contained in the Watercourses Act.[48] According to section 3 of this Act, anyone is normally allowed to use any small vessel which is not mechanically powered on common lakes and watercourses. Other boating, i.e. all boating with any significant polluting effect, may be permitted by the municipal authority under section 13 on public lakes and watercourses. If so, a set of by-laws for the area of water shall incorporate such boating provisions as the Environment Agency deems necessary. In this context regulations may be laid down to provide protection against pollution, *inter alia* provisions to restrict or completely ban the use of motorboats on certain stretches. To our knowledge, there are no more specific regulations for the prevention of water pollution.

It should also be mentioned that under section 14 of the Environmental Protection Act orders or prohibitions may be issued for the protection of water supply interests (see 3.1.2.1) and it would, for instance, be possible, under this provision, to completely ban sailing on a lake or watercourse.

Notes

1. For further details, see Ministry of the Environment circular of 22 February 1977.
2. Until 1 January 1980 the legal basis was Statutory Notice No. 524 of 26 September 1973, sections 6–8. Planning after 1 January 1980 takes place under Act No. 229 of 8 June 1978, sections 10–17. As regards the planning of water extraction and supply under the new Act, see for more details Notice No. 2 of 4 January 1980 and Ministry of the Environment circular of 25 February 1980.
3. Notice No. 162 of 29 April 1980.
4. Notice No. 6 of 4 January 1980.
5. Respectively council directive 75/440 and a directive which was approved in principle by the council on 19 December 1978. Concerning the implementation of council directive 76/160 relating to the quality of bathing water, see 4.1.8 below.
6. For further details, see the Water Supply Act, 1978, sections 18–21.
7. Some textual amendments were made to sections 12–14 of the Environmental Protection Act by Act No. 304 of 8 June 1978 as a result of the new Water Supply Act passed two days earlier.
8. In this connection, see 3.1.2.3 below.
9. See Environment Agency circular letters of 21 March 1978, 3 May 1977 and 10 April 1980. In regard to the digging up of oil tanks in the winter, see Environment Agency circular letter of 14 August 1975 and *Nyt fra miljøstyrelsen*, No. 5/1976, p. 28.
10. Notice No. 170 of 29 March 1974 with subsequent amendments.
11. With effect from 1 October 1980, section 7 of the Environmental Protection Act will be revoked by Act No. 212 of 23 May 1979 concerning chemical substances and products and replaced by the authorisations in sections 30–32 of the latter Act. In regard to the Act relating to chemical substances and product and the demarcation in this study between regulations on products and regulations concerning water pollution, refer to Chapter 10 below.
12. See section 4, subsection 2, point b, of the Notice. A parallel clause was present in the Soakaway Notice previously in force, No. 172 of 29 March 1974, section 2, subsection 1.
13. For further details on the drainage of water from roads, see Environment Agency circular letter of 22 July 1976.
14. Notice No. 174 of 29 March 1974.
15. Circular of 17 April 1974.

NOTES

16. For further details on these regulations, see Claus Tønnesen, in *Dansk Miljøret*, vol. 3, 1977 pp. 103–07.
17. See section 37 of the Sewage Notice and Sewage Circular II, D, point 11.
18. Notice No. 139 of 15 April 1980, concerning the discharge of waste-water into the ground. This Notice takes effect as from 1 June 1980, replacing Notice No. 172 of 29 March 1974 and part 12 of the Sewage Notice. The main effect of the new notice is that a considerable number of cases concerning discharge into the ground which could, under the previous regulations, only be decided by the Environment Agency and, if necessary, the Ministry of the Environment too, can now be decided in the first instance by the municipal authority or the county council. The competence of the county council especially has been extended but the municipal authority has also been given wider powers. The new notice is largely based on the Environment Agency's December 1978 report concerning 'waste-water disposal outside built-up areas'. For further details on waste-water discharge into the ground, see Ministry of the Environment circular No. 73 of 2 April 1980.
19. The normal distance requirement is 300 m but this may, with favourable hydrogeological conditions, be reduced to 75 m. It should also be mentioned that the municipal authority cannot grant permission until the medical officer's opinion has been obtained but the recommendation of the medical officer is not, as under the previous regulations, a condition of issuing a permit.
20. See 2.1.2 above concerning Chapter 5 of the Environmental Protection Act.
21. The relevant regulations were previously contained in part 12 of the Sewage Notice.
22. Notice No. 176 of 29 March 1976, see 2.1.2 above.
23. See section 7, subsection 2, of the Approval Notice.
24. For further details see 2.1.2.2 above.
25. According to Svend Andersen in *U* 1975 B 369 ff., Claus Haagen Jensen, in *Dansk Miljøret*, vol. 3, 1977, p. 62 ff., and Claus Tønnesen, *op.cit.*, p. 109 ff. The opposite view is put forward by Erik Mohr Mersing in *U* 1975 B 83 ff.
26. In a communication from the Environmental Appeal Board dated 4 November 1976 (*KFE* 1977, p. 52), it is assumed, to be sure, that orders concerning waste-water discharge from a mink-food factory could be issued under section 44 of the Environmental Protection Act. However, the undertaking had not previously been approved under Chapter 5 and it therefore made no difference, in reality, whether section 44 or section 25 was invoked as providing authority. In a later decision of 22 February 1978 (*KFE* 1979, p. 5) relating to an order prohibiting another unapproved scheduled undertaking from discharging waste-water into the ground, the Environmental Appeal Board, when dismissing an appeal against the Ministry of the Environment's decision, changed the authority invoked from section 44 to section 11, subsection 3.
27. In this connection see Environmental Appeal Board decision of 22 February 1978, mentioned above in note 21.
28. In regard to aquaculture, in particular, the Environment Agency has issued special guidelines, No. 1/1975, see especially section II thereof concerning directives for discharge, etc.
29. See pp. 8 and 24 ff. of the committee's report.
30. See for further details Notice No. 171 of 29 March 1974, section 3, subsection 5, section 9, subsection 2, and sections 11, 13 and 14.
31. In conjunction with the notice, there are also recommended standards in part 6 of the Sewage Circular and point 5 of the Sewage Guidelines. The latter chiefly concerns sampling and methods of analysis.
32. See Environment Agency report, *Miljøreformen*, November 1979, p. 119.
33. See Environment Agency report, *Miljøreformen*, November 1979, p. 34.
34. See *op.cit.*, p. 28 ff. compared with p. 35.
35. For more details on the rather unclearly drafted section 76 of the Environmental Protection Act concerning which decisions may be appealed against to the Environmental Appeal Board, see Claus Haagen Jensen, *op.cit.*, p. 137 ff.

36. See 1.7.6 above.
37. See especially *U* 1958 365 H concerning damage through leakage of phenol from a tar factory.
38. According to B. Gomard in *U* 1978 B 68 ff., in an article on liability in damages for environmental damage caused by buried chemical waste.
39. Jens Christensen, *Juristen og Økonomen*, 1977, p. 461, does not think that a legal rule concerning objective liability would produce any changes worth mentioning in practice. In regard to illegal discharges, this assessment is undoubtedly correct but otherwise unlikely.
40. For further details, see Jens Christensen, *op.cit.*, p. 459.
41. *Re U* 1969 923, see for more details Stig Jørgensen, in *U* 1970 B 192 ff. and K. Skovgaard-Sørensen, in *U* 1971 B 36.
42. Some of these rulings are referred to by Jens Christensen, *op.cit.*, p. 456 ff.
43. B. Gomard, *op.cit.*, pp. 64–66.
44. In this connection, for further details of the polluter's obligation to provide information under the Environmental Protection Act and section 52 thereof in particular, see Claus Haagen Jensen, *op.cit.*, pp. 114–21.
45. B. Gomard, *op.cit.*, p. 61, regards compensation under sections 13 and 14 as equitable compensation and not expropriation compensation. In this connection, see also Claus Haagen Jensen, *op.cit.*, pp. 158–61.
46. Statutory Notice No. 336 of 31 August 1965, with subsequent amendments. With effect from 1 July 1980, the Ship Supervisory Act is replaced by Act No. 98 of 12 March 1980 concerning ship safety, etc. As far as the present study is concerned, this change in the law does not directly prompt any additions.
47. Notice No. 173 of 21 May 1965, with subsequent amendments.
48. Statutory Notice No. 523 of 26 September 1973, with subsequent amendments.

4
Pollution of the Sea by Substances other than Oil

The current Danish legal rules on marine pollution are to be found chiefly in three acts: the Environmental Protection Act 1973 with subsequent amendments, the Dumping Act 1972 with subsequent amendments and the Oil Pollution Act of 1956 as amended subsequently. The Environmental Protection Act controls pollution of the external environment in general and contains regulations concerning pollution of surface waters, including the sea, in Chapter 4. The *Dumping Act* relates to the disposal of substances or material by throwing, emptying or sinking them in the sea from vessels, aircraft and platforms, whether fixed or floating. The *Oil Pollution Act* relates, as indicated by its name, to pollution of the sea by oil.

In relation to the Environmental Protection Act, the two latter acts may be described as 'special', meaning that, in line with general principles of interpretation, they supersede the Environmental Protection Act where its regulations govern the same circumstances.[1] Consequently, land-based pollution of the sea must be judged according to the Environmental Protection Act which must also, in principle, be invoked when pollutants other than oil are deposited in the sea or when substances are deposited there but not with a view to disposal. For instance, matters relating to the spraying of oil dispersal agents from ships are dealt with under Chapter 4 of the Environmental Protection Act.

However, the rules concerning marine pollution are undergoing rapid development and expansion and there are two new acts covering pollution of the sea which have been adopted but have not yet come into force. The *Baltic Sea Act* of 1975,[2] based on the 1974 Baltic Sea Convention, establishes regulations for the protection of the marine environment in the Baltic Sea area, which supplement the Dumping Act and the Oil Pollution Act with more stringent provisions on a number of points.

Still more far-reaching are the contents of Act No. 130 of 9 April 1980 concerning *protection of the marine environment*. This Act is designed primarily to implement through a national law the MARPOL Convention of 1973 with the corresponding protocol of 1978. The Act should also enable the provisions of the 1974 Paris Convention to be implemented; they relate to the discharge of substances originating from drilling operations. Finally, the object of the Act is to collect together all the rules relating to marine pollution from vessels, platforms, aircraft, etc. and when the Act comes into force, the previous acts relating to dumping, oil pollution and the Baltic Sea area will be simultaneously rescinded.

Both of these Acts may be brought into force by the Minister of the Environment and this will be done soon after the Baltic Sea and MARPOL conventions, respectively, come into operation. The Minister is also able, for the Marine Environment Protection Act, to decide when parts of the Act will come into force and for which geographically defined areas it will come into force. This means, among other things, that the Minister has powers to bring into effect those parts of the Marine Environment Act which correspond to the implementation in national law of the Baltic Sea Convention which has now been signed by all the states participating. This limited implementation of the Act is expected to take place at the end of 1980 and hence the Baltic Sea Act will never have any real legal effect. As no date has yet been set for the Marine Environment Act or parts thereof to come into effect, I felt obliged to base the exposition in Chapters 4 and 5 on the previous legislation. However, at some points provisions from the acts not yet in force—the Marine Environment Act, especially—will nevertheless be incorporated.

4.1 COASTAL WATERS

4.1.1 Definition of limits

The regulations concerning the demarcation of Danish territorial waters appear in a Royal Decree of 1966.[3] Under this the territorial waters are divided up in accordance with the general rules of international law into inner and outer territorial waters. The inner territorial waters consist of areas such as harbours, bays, fjords and sounds which are situated within certain more specific baselines. The outer territorial waters comprise those areas of the sea limited inshore by the baselines, the outer boundary being formed by lines drawn in such a way that the distance from any point on them to the nearest point on the baselines is 3 nautical miles.

Special provisions apply to the demarcation of fishing waters. Under a 1976 act[4] the Minister of State is authorised to establish Danish fisheries up to a width of 200 nautical miles from the baselines obtaining at any time. Where the coasts of foreign states lie less than 400 nautical miles from the Danish coast, the maximum fishing limit applicable, however, corresponds to a line running equidistant from the nearest points on the baselines off the coasts of the two states. The Minister of State has availed himself of these powers in several notices.[5]

In some environmental protection regulations, e.g. section 55 of the Environmental Protection Act, the term 'the coastal parts of the territorial waters' is used.[6] The more precise meaning of this term is not defined in either the preparatory works for the Act or in regulations issued administratively. It must be assumed that 'the coastal parts of the territorial waters' include the inner territorial waters at any rate. In addition, the term presumably covers parts of the outer territorial waters that vary geographically. Discretionary limits have to be established for these in each individual area and are presumably determined chiefly by whether the area is exposed to substantial pollution as a result of pollutants originating from sources on land.

4.1.2 Control by means of ordinary planning legislation

Readers are referred, in general, to 3.1.1 and 2.1.1 concerning general planning legislation and its connection with environmental quality planning.

As regards the coastal waters, in particular, it should be noted that environmental quality planning is still in its infancy. In accordance with a notice concerning water used for swimming and bathing beaches, bathing areas are being established in all regions.[7] Beyond this, recipient quality planning has been carried out in respect of only a few areas and it is very limited in scope. In the autumn of 1980 the Environment Agency nevertheless expects to issue guidelines for environmental quality planning for the coastal areas and it is anticipated that this will bring about a substantial stepping up of planning activities.

4.1.3 Preventive controls

In this connection readers are referred to the account above under 3.1.2. When the Marine Environment Act comes into force—see the introduction to Section 4 on the subject—the preventive rules will be considerably expanded or capable of expansion by means of regulations laid down administratively.[8] Among the preventive measures we may also include the regulations concerning the obligation to provide installations in ports to receive *inter alia* noxious fluids, sewage and refuse.[9] See too 5.1.1 below about requirements concerning the fitting out of vessels in pursuance of the Shipping Supervision Act.

4.1.4 Pretreatment

There are at present no general regulations applicable to the treatment of waste-water from land before it is discharged into coastal waters. See also 3.1.1 above.

It should be mentioned that Chapter 5 of the Marine Environment Act not yet in force lays down regulations for the pretreatment of sewage from vessels, etc. which is to be discharged less than 12 nautical miles from the coast.

4.1.5 Control of the composition and amount, etc. of discharges

Insofar as marine pollution from sources on land is concerned, readers are referred to the exposition in 3.1.4 above.

Discharges, etc. from vessels, aeroplanes and platforms are, in principle, covered by the ban in section 17 of the Environmental Protection Act on depositing polluting substances in the sea, among other places, and by the permit scheme laid down in sections 17 and 18.[10] These rules are not, however, enforced in practice in regard to the discharge of waste-water and the emptying of refuse. Some doubt may also exist as to whether the Environmental Protection Act contains the necessary clear legal authority to permit intervention in respect of traditionally acceptable discharges from vessels. However, the Marine Environment Act contains express regulations governing a number of forms of marine pollution from vessels, aeroplanes and platforms, apart even from dumping and the discharge of oil. These regulations concern, *inter alia*, fluids being transported in bulk, sewage and refuse, etc.[11] This Act also considerably extends the authorities' powers of intervention, among other things by enabling the Minister of the Environment to forbid a vessel to continue sailing or other operations if necessary in order to prevent or combat pollution of the sea.[12]

Finally, it should be mentioned that under section 4, subsection 1, points 1 and 2, of the Saltwater Fisheries Act, the disposal of articles, materials and waste products and the discharge of chemicals, mud and water containing chalk and lime deposits requires permission from the Environment Agency if these measures may be prejudicial to fishing. These regulations which may be revoked administratively by the Minister of the Environment under section 80, subsection 2, of the Environmental Protection Act apply to both jettisoning from vessels and discharges on the sea floor via pipelines.[13]

4.1.6 Supervision on the polluter's part

For the current regulations, see 3.1.6. In addition, section 19 of the Act relating to protection of the marine environment contains provisions to the effect that a journal must be kept on chemical tankers for fluids that are classed as dangerous. When this Act takes effect, it will entail an extension of the polluter's obligation to report to the environmental authorities. Thus under section 40 of the Act owners, users and captains of vessels and owners and users of platforms and the person responsible for the platform must report any disposal that has taken place contrary to Chapters 2–6 and section 58.[14]

4.1.7 Enforcement

The county council is primarily responsible for the supervision of 'the coastal areas of the territorial waters'.[15] This applies to both the state of pollution and the observance of the quality requirements set out under section 61 of the Environmental Protection Act. As regards the quality of water used for swimming, however, supervision is carried out by the municipal authority concerned.[16]

It must be assumed, however, that the municipal councils' supervision is directed in the first instance against pollution from sources on land. Insofar as pollution from vessels and platforms, etc. is concerned, the supervision is exercised by the Environment Agency, which now has 8 specially equipped craft available and will soon have another 2, and the defence and fisheries control authorities. No systematic inspection takes place, however. As regards supervision by the defence authorities especially, it has been arranged with them that the defence authorities aircraft should report cases of pollution and the polluters to the Environment Agency in conjunction with routine and training flights.[17] Nevertheless this surveillance is particularly important for oil pollution, see Section 5 below. It should be added that the Marine Environment Act will involve an official obligation subject to penalty for the captains of vessels and aircraft to report sightings of oil spills and harmful floating substances.[18] This Act also empowers the supervisory authorities to undertake the inspection of vessels and platforms. This will be feasible, except for foreign vessels passing through Danish territorial waters, even in the absence of a well-founded suspicion that the provisions of the Act have been violated.[19]

In regard to the possible imposition of *penalties* and penal practice, see section 3.1.7.

4.1.8 Quality targets for coastal waters

4.1.8.1 LEGAL REQUIREMENTS AND RECOMMENDED GUIDELINES

Binding provisions concerning the quality of coastal waters exist only in regard to water for bathing. In this sector the Minister of the Environment has, under section 4 of the Environmental Protection Act established regulations in a notice for both fresh water and salt water, implemented EEC council directive 76/160 in Danish law.[20] Under these regulations, the bathing water must not for more than 5% of the time in the bathing season contain more than 10,000 coliform bacteria per 100 ml and not more than 1000 *E.coli* per 100 ml. Furthermore, the bathing water must not be discoloured, have a visible film of mineral oil or permanent scum and must not smell of mineral oil and phenols. The county council may, however, lay down stricter requirements for the quality of bathing water than those established by the general regulations.

In regard to the recommended standards contained in the Environment Agency's Guidelines No. 6/1974 concerning waste-water, see section 3.1.8.1.

Finally, quality requirements are also being laid down regionally concerning coastal waters too as an element in environmental quality planning. See 4.1.2 above in this connection.

4.1.8.2 INFORMATION CONCERNING OBSERVANCE OF THE TARGETS

Readers are referred in general to the Environment Agency's report *Miljøreformen* of November 1979, pp. 152–61. As regards bathing water on the coast, information is collected annually from the counties and the Greater Copenhagen County and on the basis of this data the Environment Agency prepares and publishes nationwide bathing water surveys each year. The information collected shows that since 1972 there has been a substantial improvement in the quality of bathing water.[21]

4.1.8.3 RESTRICTIONS ON THE USE OF SEAWATER THAT DOES NOT MEET THE TARGETS

If bathing water does not comply with the above-mentioned quality requirements, the municipal council must, under section 3 of the Bathing Water Notice, prohibit bathing unless reasonable conditions can be attained within a time limit fixed by the municipal authority.

4.1.9 Individuals' rights

By decision of the Environmental Appeal Board of 24 March 1980 mentioned above under 1.7.3.2, it was assumed that neither the Danish Society for the Preservation of Natural Amenities nor the Danish Anglers' Federation was entitled to appeal against a decision on permission to discharge waste-water into the Little Belt strait. This ruling has quite far-reaching consequences. First, it must be assumed that the two above-mentioned organisations, which are the two associations most obviously entitled to appeal about decisions relating to discharge into the sea, will not be considered entitled to appeal in other similar cases. Second, the ruling makes it unlikely that the right of individuals to appeal about discharges into the sea will be recognised apart from very special situations.[22] In other words, in cases of this kind virtually no-one will be entitled to appeal other than polluter and the public authorities specially designated in sections 74 and 80 of the Environmental Protection Act, since other individuals and associations are not assumed to have an 'individual, essential interest in the outcome of the case'. Third, this means that anyone other than the polluter will not have the status of an interested party under the Public Access to Documents Act with the greater access to information and documents conferred thereby, the right to have settlement of a case deferred until an opinion has been submitted and the right to bring a civil action in the courts concerning the administrative authorities' decisions.

In regard to individuals' rights, readers are also referred in general to 3.1.9 with further references to Section 1. It should be added as far as the *right to appeal* is concerned, that the Marine Environment Protection Act will extend the jurisdiction of the Environmental Appeal Board in marine pollution cases. Under section 53 of this Act it will be possible to appeal to the Environmental Appeal Board against decisions on *inter alia* permits for burning off at sea and intervention in the form of orders and prohibitions. It is anticipated, however, that there will in practice be very few appeals to the Board as a result of these new rules.

Questions concerning *liability in damages* for injury caused by the above-mentioned forms of marine pollution are unlikely to arise. Nevertheless, the question may occasionally arise—even apart from cases concerning oil pollution—of damages for measures taken in respect of marine pollution through actions contrary to the regulations or to prevent a definite danger of this. Under the current regulations, claims for damages have to be dealt with on the basis of the general culpa rule, for further details see 3.1.9 above. If it is a question of the authorities' expenditure on combating pollution or preventing a definite danger of it, the Marine Environment Protection Act imposes a stricter basis of liability, as under section 46 of the Act, such costs must be unconditionally reimbursed by the owner of the vessel or platform if they may be regarded as 'reasonable'.[23]

It should further be noted that compensation for disposal and discharge which are permitted by the Environment Agency under section 4 of the Saltwater Fisheries Act and which cause damage to the fishing industry must be paid on the lines of the expropriation theory.

4.2 CONTROL OF DUMPING FROM SHIPS AND AIRCRAFT

Since 1972 the right to effect dumping at sea from vessels and aircraft has been subject to legal regulation. This was originally done by means of a notice which was issued in accordance with a clause incorporated in the Oil Pollution Act in 1971.[24] In 1972 a separate act was passed 'relating to measures against marine pollution by substances other than oil' which came into force on 7 April 1974.[25] This Act is predominantly concerned with dumping at sea from vessels, aircraft and platforms and is usually referred to as the *'Dumping Act'*.[26] The Act was amended in some respects in 1975.[27] However, a number of these amendments were made with the implementation of the Baltic Sea Convention regulations in mind and have not yet come into force. The sector of application of the Dumping Act is covered by the Marine Environment Act 1980, in particular Chapter 8 thereof, and the former act is expected to be repealed when parts of the Marine Environment Act take effect around the end of 1980/early 1981.[28]

The current regulations in the Dumping Act and the corresponding provisions in the Marine Environment Act are based on the London Convention of 1972 concerning the prevention of marine pollution through the dumping of refuse and other substances, the Oslo Convention of 1972 concerning the prevention of marine pollution through dumping from ships and aircraft, the Baltic Sea Convention of 1974 and the Paris Convention of 1974 on the prevention of land-based marine pollution.

4.2.1 Geographical extent of Danish jurisdiction and substantive rules

The Dumping Act covers, on the one hand, means of transport whose owners are Danish (or Faeroese) and which are on or over Danish territorial waters (including territorial waters round the Faeroe Islands) and, on the other hand, platforms with Danish (or Faeroese) owners or which are situated in Danish territorial waters or on the Danish continental shelf area (including territorial waters and the continental shelf area around the Faeroe Islands). In this connection, see 4.1.1 above concerning the demarcation of Danish territory at sea. The corresponding regulations in section 2 of the Marine Environment Act are drafted differently due to the wider scope of the Act but there are no major differences in regard to dumping.

There is under the Act a *ban on all dumping* of substances or materials in Danish territorial waters (including territorial waters around the Faeroe Islands), in northern parts of the Atlantic Ocean as defined under the Oslo Convention and in the Baltic Sea, Øresund and the Belts (straits). In other sea areas the *dumping of certain substances or materials is banned*. These substances are listed in Appendices 1 and 2 to the Act which may be amended by the Minister of the Environment and, in fact, were with effect from 1 May 1980.[29] The contents of the appendices are in accordance with the Oslo and London Conventions. The bans do not, however, cover dumping that takes place in order to save human lives or with a view to protecting the safety of the means of transport (or platform).

These bans may be waived by permits. The decision to grant a permit is normally taken by the Environment Agency but in a 1979 notice it was stipulated, with effect from 1 May 1980, that permission may also be granted by the competent authority in a country which signed the London Convention.[30] The Act sets out various conditions for the issue of dumping permits. Permission to dump substances and materials covered by Appendix 1 (the 'black' list) may be given only in concrete form and only in quite special circumstances. Permits for the dumping of substances and materials covered by Appendix 2 (the 'grey' list) may also only be granted in specific individual cases but no further material conditions are imposed in the Act. Permits for the dumping of substances and materials not covered by Appendices 1 and 2 and refuse containing only a negligible amount of substances and materials mentioned in Appendix 2 may be issued without restriction as to time or to the number or extent of dumping operations. In regard to the Baltic Sea area a regulation that has not yet come into force was adopted in 1975 to the effect that material removed from the sea bottom may only be dumped by depositing it elsewhere on the sea bottom ('hopper discharge') always provided that the material does not contain substantial amounts or levels of substances set out in a new Appendix 3. The ban on dumping at sea is extended in the Act by a ban on handing over substances or materials for transport or transporting and loading them with a view to dumping, unless a dumping permit exists, see Chapter 7.

The Marine Environment Act will not give rise to any major changes in the rules concerning prohibitions and permits. It should nevertheless be mentioned that under this Act any dumping in all sea areas will require permission and the Minister of the Environment may decide that permits for 'hopper discharge' may be issued by the county council and in the metropolitan area by the Greater Copenhagen County Council.

4.2.2 Supervision

Supervision of the observance of the Dumping Act and of regulations issued in pursuance thereof is carried out under section 9 of the Act by the police, the defence authorities and the fishery control authority. The Environment Agency also exercises a certain amount of supervision, having at its disposal eight special vessels, and is expressly named as the control authority in section 47 of the Marine Environment Act. See also 4.1.7 on supervision.

4.2.3 Enforcement

Legal requirements founded on or in pursuance of the Dumping Act are reinforced by *penalties* in the same way as the provisions of the Environmental Protection Act, see 2.1.6 above in this connection. To the best of my knowledge there are no examples of criminal cases brought under the Dumping Act.

In the event of well-grounded suspicions that the regulations are being violated, the supervisory authorities named in the Dumping Act may inspect a vessel (or a platform), see section 9, subsections 2 and 3, of the Act. In this respect the Marine Environment Act will substantially extend the authorities' powers. On the one hand, it will be possible for vessels (and platforms) to be inspected by the supervisory authorities without well-founded suspicion that the regulations have been violated, except when foreign vessels (and platforms) passing through Danish territorial waters are not involved, see section 47, subsections 2 and 3. On the other hand, these authorities can undertake investigations aboard ships and platforms and prohibit further sailing, etc., see sections 44 and 45.

4.3 CONTROL OF POLLUTION FROM EXPLOITATION OF THE SEA OR THE SEA BED

4.3.1 Geographical extent of Danish jurisdiction and substantive rules

In regard to the extent of Danish territory at sea, readers are referred to 4.1.1. The Danish sector of the continental shelf was established in accordance with the Convention of 29 April 1958 relating to the continental shelf, as its provisions were defined for the North Sea area in the International Court's decision of 20 February 1969.[31]

Marine pollution resulting from the exploitation of the sea or sea bed is chiefly regulated in the same way as marine pollution caused otherwise. In regard to territorial waters, this follows from the individual environmental acts; in regard to the Danish sector of the continental shelf, the Continental Shelf Act, section 3, states that Danish law applies to installations situated in the shelf area for the purpose of exploring or exploiting the continental shelf and also in the safety zones around the installations, see section 4, subsection 2, of the same Act.[32] This means that oil pollution is assessed under the Oil Pollution Act (see Chapter 5), that dumping from platforms and the like is governed by the rules in the Dumping Act, see 4.2 above, and that other pollution matters must at present be dealt with under the Environmental Protection Act and regulations issued in pursuance thereof, see 4.1 above with references to Chapter 3. In this context it should also be noted that the Marine Environment Act 1980 will substantially extend and give more force to the regulations on preventing and combating pollution from platforms installed for exploiting raw materials in the sea bed, both in territorial waters and on the continental shelf, see 4.1 above in particular.

Of the special regulations in this sector, particular mention should be made of the permit schemes set up for exploitation of the sea bed and its underground mineral and other inanimate resources. The exploitation of such raw materials, which were not exploited by the private sector in this country until 23 February 1932—for instance, oil, natural gas and salt—requires a licence from the Minister of Energy under the Underground Resources Act.[33] The exploitation of other raw materials, such as stone, gravel, sand, lime and chalk, may under the Raw Materials Act, section 15, take place only by permission of the Minister of the Environment.[34] The Continental Shelf Act moreover, in its section 4, authorises the Minister of Energy to lay down rules concerning *inter alia* measures for the prevention or remedying of pollution. Such regulations have not yet been issued, however.

4.3.2 Supervision

As regards supervision of compliance with the general pollution regulations, see 4.1.7, 4.2.2 and 5.1.5. In respect of the permit scheme under the Raw Materials Act, the Minister of the Environment can lay down regulations for the surveillance of raw material extraction in territorial waters, see section 20, subsection 1, of the Act, but no such rules have been laid down as yet. In the absence of express provisions concerning surveillance, it has to be carried out by the authority competent to issue permits, i.e. the Ministry of the Environment. The Underground Resources Act contains no surveillance provisions but in regard to operations exploiting underground resources covered by this Act, there is in practice no need to supervise observance of the obligation to apply for a permit.

4.3.3 Enforcement

In regard to the enforcement of the general pollution regulations, see 4.1.7, 4.2.2 and 5.1.5. Concerning the permit schemes under the Underground Resources Act and the Raw Materials Act in conjunction with the regulations of the Continental Shelf Act, it should be noted especially that there are rules for the withdrawal of licences and permits and that to disregard these schemes generally involves liability to punishment.[35]

Notes

1. See also, therefore, Niels Borre, *Miljøbeskyttelsesloven* ('Environmental Protection Act'), 1973, p. 90 ff.
2. Act No. 324 of 26 June 1975.
3. Decree No. 437 of 21 December 1966, as amended by Decree No. 189 of 19 April 1978.
4. Act No. 597 of 17 December 1976.
5. Notice No. 639 of 22 December 1976 (the North Sea), Notice No. 639 of 22 December 1977 (Kattegat and Skagerrak) and Notice No. 43 of 1 February 1978 (Great Belt and the Sound).
6. See too, Notice No. 177 of 29 March 1974, section 7, subsection 1.
7. Notice No. 143 of 30 March 1978.
8. See, for example, the Marine Environment Protection Act, sections 17 and 18, 21, 32 and 34.
9. In pursuance of section 20 of the Baltic Sea Act, the Minister of the Environment laid down regulations concerning receiving arrangements in Notice No. 379 of 27 July 1978 which cannot, however, take effect until the Act has come into force. In a circular linked to the Notice—No. 128 of 27 July 1978—the Minister nevertheless emphasised to Danish port authorities and municipal councils that the work of planning receiving arrangements must be started as soon as possible if the receiving installations were to be available at the latest 1 year after the Baltic Sea Convention takes effect generally.
10. The order regulations in section 25 of the Environmental Protection Act will, by their wording, also be applicable in principle.
11. Chapters 3, 5 and 6 of the Act together with section 33 relating to burning off at sea.
12. Section 45 of the Act.
13. In this connection, concerning pipelines in territorial waters, see the Minister of Transport's Circular No. 101 of 18 May 1977.
14. Compare this with the more restricted rule concerning the reporting obligation in the Baltic Sea Act, section 23.
15. Section 55 of the Environmental Protection Act and Notice No. 177 of 29 March 1974, section 7.
16. Notice No. 143 of 30 March 1978 concerning bathing water and bathing beaches, section 5.
17. In this connection, see report of 29 February 1980 by the Danish parliament's Environment Committee, pp. 5 and 6, on considerations relating to the systematic surveillance of Danish waters from the air.
18. Section 40, subsection 2, of the Act.
19. Section 47 of the Act.

20. Notice No. 143 of 30 March 1978, the 'Bathing Water Notice'.
21. See above-mentioned report of the Environment Agency, p. 161 ff.
22. As such an exceptional case, we may perhaps imagine landowners immediately on the coast appealing against a ban on bathing, see 4.1.8.3 above.
23. See in this connection the letter from the Ministry of Justice dated 18 March 1980 printed in the supplementary report of 19 March 1980 by the Danish parliament's environment committee, pp. 4–6. It is assumed there that the compensation rule in section 46 of the Act, in accordance with the convention on which it is based, covers not only measures on account of pollution that has already occurred but also precautions to avert a definite danger of pollution contrary to the provisions of the Act.
24. Notice No. 19 of 18 January 1972.
25. Act No. 290 of 7 June 1972, brought into effect by Notice No. 156 of 25 March 1974.
26. Section 11 of the Act authorises the Minister of the Environment to issue bans on the jettisoning and emptying of substances and materials originating from the normal operation of ships and aircraft, etc. These powers have not yet been used, however.
27. Act No. 312 of 26 June 1975.
28. In this connection, see the introduction to Chapter 4.
29. Notice No. 552 of 20 December 1979.
30. See Notice No. 156 of 25 March 1974 and Notice No. 552 of 20 December 1979.
31. In this connection, see Royal Decree No. 259 of 7 June 1963.
32. Statutory Notice No. 182 of 1 May 1979.
33. Act No. 181 of 8 May 1950; see Continental Shelf Act, section 2, subsection 1.
34. Act No. 237 of 8 June 1977.
35. See in regard to penalties, section 36 of the Raw Materials Act and section 5 of the Continental Shelf Act. The Underground Resources Act contains no penal clauses but, under section 5, subsection 2, of the Continental Shelf Act, contraventions of conditions in permits and licences under both the Raw Materials Act and the Underground Resources Act are always punishable by a fine.

5
Pollution of the Sea by Oil

5.1 CONTROL OF DISCHARGE FROM SHIPS

5.1.1 The fitting out and equipment of a ship

The regulations covering the fitting out and equipment of ships afford some guarantee that the sea will not be polluted by oil from vessels. The great majority of Danish vessels are, in accordance with the *Shipping Supervisory Act*,[1] approved by the State Shipping Inspectorate, which is a directorate under the Ministry of Industry. At the time of approval ships must comply with a number of more detailed regulations that are set out in Notice No. 173 of 21 May 1965 with subsequent amendments. These provisions have hitherto been designed more to ensure the safety of crew and vessel but some regulations have also counted as anti-pollution measures. Act No. 279 of 26 June 1975 amended the Shipping Supervisory Act following the signing by Denmark of the Baltic Sea Convention. Under the heading of 'environmental protection' a new clause was incorporated under which ships must be built and equipped in such a way that as much consideration as reasonably possible is given to protecting the sea from pollution through the discharge of oil, noxious substances, sewage and refuse. However, the 1975 Act of amendment has no more been put into effect than the Act relating to protection of the marine environment in the Baltic Sea, in which connection see the introduction to Chapter 4. With effect from 1 July 1980, the Shipping Supervisory Act is entirely replaced by Act No. 98 of 12 March 1980 concerning the safety, etc. of ships. Section 2 of this Act contains regulations concerning the building and equipment of vessels which correspond, in regard to the prevention of marine pollution, with the as yet unimplemented provisions of the Shipping Supervisory Act mentioned above. When regulations are issued by notice in pursuance of section 2, subsection 5, of the new Act, it will be possible for the requirements of the MARPOL Convention concerning the fitting out and equipment of ships to be fulfilled.

It should further be noted that under section 6 of the *Oil Pollution Act* Danish vessels—tankers of 150 registered tons gross and over and other vessels of 500 registered tons gross and over—must be fitted out in such a way that oil cannot escape into the gutters if their contents are emptied into the sea without having passed through a separator for oily water.[2]

5.1.2 Ship's crew

General rules concerning ships' crews are contained in the *Crew Act*.[3] However, none of its provisions are specially concerned with the prevention of marine pollution by oil.

5.1.3 Emptying of oil

The right to empty oil into the sea has been regulated by law in Denmark since 1956 when the first Act was passed concerning pollution of the sea by oil.[4] For obvious reasons, the regulation of this and also of most other aspects of marine pollution has followed the international agreements in this sector and in regard to oil pollution this is chiefly the 1954 international convention on oil pollution of the sea with subsequent amendments. The convention amendments have given rise to some changes in Danish legislation, always in the direction of tightening it up. After an important legal amendment dating from 1971 was brought into effect in 1978,[5] with the aim of implementing a convention amendment of October 1969, the Ministry of the Environment collected the current legal regulations concerning oil pollution in Statutory Notice No. 134 of 28 March 1978.

A characteristic of the *current regulations* is that they relate solely to 'heavy' oils and mixtures containing heavy oil.[6] For these types of oil the Act lays down a ban on 'emptying' by which is understood 'any form of emptying or discharge, irrespective of cause'. In *Danish territorial waters* (see 4.1.1) it is absolutely prohibited to empty oil of the kind mentioned, see section 2 of the Act. In regard to *other sea areas* the Act substantially restricts the rights of *Danish vessels* to empty oil. Section 3 contains conditions for the discharge of oil by tankers of 150 register tons gross and over. These include the requirement that the ship is more than 50 nautical miles from the nearest baseline. Section 4 sets out conditions for other vessels of 500 registered tons gross and over to be able to empty oil. The restrictions in sections 3 and 4 do not apply, however, if the discharge takes place on account of the vessel's safety, in order to prevent damage to ship or cargo or to save human lives or if the discharge is due to damage to the vessel or unavoidable leakage, provided that reasonable steps are taken to combat it. The regulations concerning discharge outside Danish territorial waters agree closely with the international convention of 1954 as subsequently amended.

With the coming into force of the *Marine Environment Act* of 1980 (see introduction to Chapter 4 in this connection) the Oil Pollution Act will be repealed and replaced by Chapter 2 of the new Act, which is drafted on the basis of the 1973 MARPOL Convention. On two points especially, the Marine Environment Act will make the regulations more stringent. First, it covers the emptying of any form of mineral oil and mixtures thereof. Second, Chapter 2 of the Act applies to all tankers and other vessels of 400 gross registered tons and over.

The above-mentioned international agreements concerning pollution of the sea by oil are supplemented by three regional treaties, of which Denmark is a signatory: the Bonn Agreement of 1969 concerning the North Sea, a Nordic agreement of 1971 and the 1974 Baltic Sea Convention. The first two treaties did not, as far as Denmark was concerned, call for implementation measures at home. However, the Baltic Sea Convention generated supplementary regulations on oil pollution in the 1975 Baltic Sea Act. As mentioned in the introduction to Chapter 4 of this study, this Act will not be implemented, however, as its provisions are covered by the Marine Environment Act of 1980.

5.1.4 Installations for receiving oil residues

Under section 8 of the Oil Pollution Act, it is for the Minister of the Environment to decide, in collaboaration with the Minister of Public Works, in which major ports installations are to be established to receive oil residues from vessels other than tankers, see Art. VIII of the 1954 Convention. More detailed provisions to this effect have not been enacted since all the commercial ports of any significance possess such installations. Oil residues from tankers are discharged at the oil refineries.

The Marine Environment Act will extend the obligation to establish receiving installations. On the one hand, this Act, as mentioned, covers all forms of mineral oil and not just heavy oils. On the other, bodies other than port-owners may be compelled to set up receiving installations, for instance ship-repair establishments and, moreover, under the Marine Environment Act, receiving installations must be set up for substances other than oil residues *inter alia* noxious fluids, sewage and refuse.[7]

5.1.5 Enforcement

5.1.5.1 SUPERVISION

The polluter is under an obligation to supervise his own operations which extends beyond that arising from his or her interest in avoiding criminal liability or liability in damages. It should be particularly mentioned in this context that all vessels covered by the regulations described under 5.1.3 concerning the discharge of oil into the sea must keep an *oil journal*[8] under the Oil Pollution Act. The ship's captain must, moreover, under section 14 of the Shipping Supervisory Act regularly inspect the ship and test all equipment upon which the safety of the vessel and of human lives depends. The obligation to test the ship's equipment can in any case scarcely be assumed to be applicable to requirements imposed in order to prevent oil pollution. The Marine Environment Act will, by reason of the provisions in section 40, subsection 1, concerning the reporting obligation quite substantially tighten up the 'polluter's' duty of supervision and his obligation to inform the authorities about existing or threatened marine pollution on his own initiative.

The polluter's surveillance is and must necessarily be supplemented by surveillance on the part of the *authorities*. The task of inspection is, under both the Oil Pollution Act and the Shipping Supervisory Act officially delegated to the Shipping Inspectorate. Under section 10 of the Oil Pollution Act the latter is entitled to inspect Danish and foreign vessels in Danish ports or in Danish territorial waters in order to investigate a definite suspicion that the provisions of the Act or of the underlying Convention have been violated. There is no arrangement for routine or systematic inspection to see that these regulations are observed. Inspection and investigations are carried out only if there are special grounds for doing so. Such grounds would exist, in particular, if reports were received from the Environment Agency, vessels, the air force, and navy or civilian aircraft of oil being discharged at sea. In this connection, see further details in 4.1.7 above concerning surveillance by the defence branch, etc. The control regulations of the Shipping Supervisory Act concern only Danish vessels in the first instance but they may be and have been extended to cover as a general rule foreign vessels, in

Danish ports and in Danish territorial waters.[9] The inspection is carried out at the time of acquisition or conversion[10] and partly through extraordinary check inspections which may be arranged at the discretion of the Shipping Inspectorate.[11] Section 21 of the Shipping Supervisory Act also makes it incumbent on the police, the Customs authorities and port authorities among others, to report infringements of the provisions of this legislation.

In this respect too the Marine Environment Act involves an extension of the regulations, in this connection see 4.1.7 above. In particular, the effect of section 40 of this Act will be to empower the authorities by and large to carry out inspections without reference to whether there is a definite suspicion of the legislation being violated and to introduce an obligation subject to penalty for, among others, the captains of all vessels and aircraft to report sightings of substantial spills of oil (or dangerous floating substances). This will legally involve *private individuals other than the polluter* in supervising compliance with the regulations on oil pollution.

5.1.5.2 COUNTERMEASURES

Infringement of the provisions of both the Oil Pollution Act and the Shipping Supervisory Act is *subject to punishment*. Violations of the first-mentioned Act are now punishable within the same framework as for the Environmental Protection Act, i.e. fines, detention and, with aggravting circumstances, imprisonment for up to one year, in this connection see 2.1.6.2 above. To our knowledge there has in the Danish courts been only one verdict (in 1979) imposing a penalty for disregarding the Oil Pollution Act. A minor discharge was involved which was reported by the Swedish authorities and the case was settled with the imposition of a fine. There have, in addition, been a number of infringements of the provisions of the Act relating to the oil journal, but these are normally settled by reprimands to those at fault. It should be added that many cases of illegal discharge have, of course, been observed where there was no case for attributing the violation to a particular vessel.

To disregard the regulations of the Shipping Supervisory Act may lead to the imposition of a fine. Yet no legal proceedings have taken place in regard to fitting-out requirements to prevent oil pollution.

If it is observed or suspected that the Act is being violated, this may also entail the application of *administrative measures*. Under section 30 of the Shipping Supervisory Act a vessel can be detained if defects in its fitting out and equipment etc. mean that there would be obvious danger to human life in allowing the vessel to sail.[12] From a practical viewpoint, however, this rule could not be applied in connection with factors that are linked with pollution of the sea by oil. On the other hand, the new provisions concerning intervention in the form of orders and bans on sailing in Chapter 12 of the Marine Environment Act will be applicable in practice in cases of serious marine pollution or the danger of it, whether the pollution is due to oil or something else.

5.1.6 Civil law liability in damages

Originally, issues concerning damages for injury due to oil pollution were basically to be judged according to the general rule of *culpa*, which was also the basis for the explicit provisions concerning liability in damages in the Marine Act in the event of collisions between vessels. An extension of this basis of liability might in some cases follow from the general principle of the law of necessity. If a vessel discharged oil in order to avert the threatened danger of a greater loss—typically the ship herself and her cargo—it would probably be impossible to impute liability under the *culpa* rule, but the party in whose interest the discharge took place would be responsible for damages in accordance with the rules of the law of necessity if the discharge was negligent unless the situation constituted an emergency. The prerequisite for law of necessity liability was that in the individual case beneficial consequences for the discharge might be demonstrated and that these outweighed the harmful ones. The law of necessity situation often existed in relation to injury from oil pollution and the legal position was therefore closer to objective liability in damages, although there was still a certain gap between them.[13]

Among other things, Act No. 227 of 24 April 1974, introduced a new Chapter 12 into the Marine Act concerning responsibility for oil damage. These new provisions accord by and large with the 1969 Convention concerning civil law liability in damages for damage caused by oil pollution and the 1971 Convention concerning the setting up of an international fund for compensation for damage caused by oil pollution. The general rules of the Act in regard to liability in damages were brought into force on 19 June 1975[14] and the provisions concerning the international compensation fund on 16 October 1978.[15]

In accordance with these the principal rule of objective liability for oil damage is clear. The owner of a vessel which transports oil in bulk as her cargo is responsible, without reference to blame, for any damage outside the ship which is due to pollution caused by oil, including bunker oil, which is discharged or escapes from the vessel and also for expense or damage caused by reasonable precautions taken after the occurrence of the event in question in order to avert or limit the said injury. In a statement made during the reading of the Marine Environment Bill in the Folketing, the Minister of Justice assumed (rightly in my opinion) that the above-mentioned compensation clause of the Marine Act and the Convention are also applicable to both expense incurred after the discharge or escape of oil and to expenditure on reasonable precautions with a view to averting a definite danger of oil pollution.[16]

The compensation rules of the Marine Act mainly follow the provisions of the international convention. On two points, however, Danish legislation tallies with a joint Nordic Bill which is more stringent than the Convention. First, objective liability in damages will also apply in cases where the oil pollution damage occurs ashore, here as a result of discharge from vessels which do not transport oil in bulk as cargo (Section 282, subsection 2). Second, the liability rules also cover warships or other vessels which are owned or used by a state and which are, at the time of discharge, being used solely for non-commerical state purposes if the injury is caused in Denmark (Section 283).

The 1974 Act has also altered the legal position in two other respects. On the one hand, the earlier provisions concerning limitation of the shipowner's liability in damages were amended in keeping with the provisions in the Liability Convention of 1969.[17] On the other, there is introduced, on the basis of section 274 of the Act, an obligation for owners of Danish vessels carrying more than 2000 tons of oil in bulk as cargo to insure themselves against liability in damages for oil damage.[18]

The question of liability for oil pollution damage has only arisen in the Danish courts in a single case (*U* 1952.53 H) when the Ministry of Defence was found responsible for damage to the fishing catch because there had been abnormal oil spillage owing to negligence on the crew's part. In recent years the environmental authorities have, moreover, claimed costs from shipowners in a number of cases for combating oil pollution. In every instance these cases were settled amicably. The clear rules concerning the basis of liability which have been applicable since 1975 have undoubtedly contributed to court proceedings not being considered necessary by any of the parties.

5.2 CONTROL OVER INSTALLATIONS ON THE COAST AND ACTIVITIES IN PORTS

5.2.1 Siting and construction of installations on the coast

A number of the installations which present the greatest danger of marine pollution by oil are covered by the *approval scheme* of the *Environmental Protection Act*. This applies to plants for the refining of mineral oil, plants for the processing or destruction of waste oil, overground storage depots containing more than 10,000 m³ of mineral oil or fluid mineral oil products and oil-fired heating installations with thermal outputs of 5 MW and above.[19] For such installations the conditions of the actual approval will set out in more detail how the installation is to be arranged in order to prevent *inter alia* pollution of the sea by oil. For further details of the approval scheme under the Environmental Protection Act, see 2.1.2 above.

5.2.2 Loading and unloading

Rules for the loading and unloading of oil are to be found in the *regulations of the individual ports*, see in this connection the standard port regulations of 19 November 1931 with subsequent amendments, sections 8 and 16–23.[20] General regulations specifically aimed at preventing marine pollution can, moreover, be issued by the Minister of Industry in purusance of section 7, subsection 1, point 2, of the *Ships' Safety Act*, which took effect on 1 July 1980.[21] Such general regulations have not yet been laid down. It should be added that for undertakings, etc. that have to be approved under Chapter 5 of the Environmental Protection Act, specific conditions may be stipulated concerning loading and unloading *inter alia*.

5.2.3 Enforcement

In this connection readers are referred to the account above under 2.1.5–6 and also 2.3.2.

5.3 CONTROL OVER FIXED INSTALLATIONS AT SEA

As far as Denmark is concerned, questions concerning pollution of the sea by oil from fixed installations at sea will, as far as can be seen, occur in practice only in connection with installations for prospecting for and recovering oil from beneath the sea floor. Installations of this type at present exist only in the North Sea.[22]

5.3.1 Extent of Danish jurisdiction

For this, see 4.1.1 and 4.3.1 above.

5.3.2 Approval schemes

Prospecting with a view to the exploitation and recovery of oil deposits in territorial waters and the Danish sector of the continental shelf requires a licence from the Minister of Energy in accordance with the Underground Resources Act of 1950.[23] In this connection, see 4.3.1.

5.3.3 Construction, equipment and safety zones

Provisions concerning the construction and equipment of the installations are laid down in the licences mentioned under 5.3.2. Under section 4, subsection 2, of the Continental Shelf Act, the Minister of Energy can draw up regulations concerning safety zones around installations which are used for prospecting and exploitation. The zones must extend for no more than 500 m around the installations in question measured from any point on the outer edge of the latter. In pursuance of this clause the Minister has established a safety zone around 'Danfeltet' in the North Sea by a 1976 Notice.[24] It should be added that the Minister of Energy can also, under section 4, subsection 1, of the same Act, lay down general regulations concerning safety measures for the erection and operation of such installations and measures for the prevention or remedying of pollution. These powers have not been used so far.

5.3.4 Manning

There are no general regulations in regard to the manning of the installations that are dealt with in this subsection. Any requirements in this connection may be established by the Minister of Labour in pursuance of the legislation on labour welfare (Working Environment Act), see especially section 41 of the Working Environment Act.

5.3.5 Discharge of oil

The *current regulations* concerning the discharge of oil from platforms are summary in the extreme. If the platform is situated in Danish territorial waters, there is under section 2 of the Oil Pollution Act a definite ban on any form of discharge or overflow of oil, regardless of cause. As mentioned previously, this, like other clauses of the Oil Pollution Act, applies only to 'heavy' oils. This Act contains no regulations covering installations situated on the Danish sector of the continental shelf outside the territorial waters but such platforms are, in principle, covered by section 17 of the Environmental Protection Act.[25] Under this regulation substances which may pollute water must not be released into the sea *inter alia* without permission. For further details, see under 3.1.4.

The *Marine Environment Act* will create more detailed and precise rules in this sector. From platforms situated outside a 'special sea area' discharge may take place according to the same regulations as for vessels of 400 gross registered tons and above, see section 13, subsection 1.[26] As these regulations assume that the platform is in motion, it means that the discharge of oil from the stationary platforms involved here is not permissible. If the platform is in a 'special sea area', to which category the Baltic Sea will belong, discharge must only take place subject to the use of approved systems for the monitoring and control of oil discharge and for separating oily water and oil filtration or similar approved systems always provided that the oil content of the discharge is less than 15 ppm without dilution. The regulations of the Marine Environment Act concerning discharge do not, however, apply to the emptying into the sea of substances or materials which result directly from the investigation or exploitation of undersea mineral deposits, see section 3, subsection 3, of the Act. For operations of this kind the Minister of the Environment is able, under section 32 of the Act, however, to establish regulations to protect the marine environment. Finally, it should be mentioned that Section 33 of the Act contains more detailed clauses concerning permission to burn off substances and materials at sea, something which is at present regulated by the Environmental Protection Act.

5.3.6 Loading and unloading

There are at the moment no general provisions concerning these operations. Presumably section 32 of the Marine Environment Act, which is referred to under 5.3.5, will contain the powers necessary for the Minister of the Environment to draw up regulations on the subject. It will, moreover, be possible for requirements to be specified as the conditions of a permit to prospect and extract.

5.4 CONTINGENCY PLANS FOR DEALING WITH OIL POLLUTION ACCIDENTS

The *current legislation* contains no directives aimed at countering the consequences of accidents which entail pollution of the sea and coasts by oil (or chemicals). Since 1970 there has, however, been a permanent national state of preparedness under the Environment Agency to combat marine pollution by oil and chemicals and this has gradually been extended. Today the Environment Agency has at its disposal 8 small special craft, including 2 specially designed to combat chemical pollution, together with a number of oil-skimmers and floating barriers. With the aid of financial grants in 1977–79, a substantial expansion in terms of equipment and staff is in progress and the Environment Agency will soon have an additional two large special craft, 2 smaller special craft and 3 oil lighters. The Environment Agency feels that this expansion will provide reasonable facilities for combating oil spills of up to 10,000 tons.[27]

Chapter 10 of the *Marine Environment Act* contains a series of clauses concerning readiness to combat oil and chemical pollution. The provisions are designed chiefly to produce a clear division of responsibility and tasks. The preparedness plan is made up as follows:

(1) state oil and chemical preparedness under the Environment Agency to combat marine pollution;

(2) readiness of the ports under the municipal authorities to combat pollution in the ports;

(3) beach-cleaning preparedness, which is the responsibility of the mu-

nicipal authority or the county council or the Environment Agency, depending on the nature and extent of the pollution, when it has not been possible to remove the latter at sea.

The Act makes the local authorities responsible for drawing up contingency plans, which are coordinated by the county council. The contingency regulations in the Marine Environment Act are expected to be supplemented by a forthcoming Act concerning hydrocarbon installations at sea. It is assumed that this will contain contingency regulations which will *inter alia* compel the licensee to establish an approved state of readiness to combat marine pollution.

Notes

1. Statutory Notice No. 336 of 31 August 1965 with subsequent amendments.
2. Further regulations on this subject are laid down by Notice No. 155 of 12 May 1959.
3. Statutory Notice No. 556 of 12 November 1975.
4. Act No. 70 of 28 March 1956.
5. Act No. 49 of 3 February 1971 which was brought into effect by Notice No. 2 of 4 January 1978.
6. For further details, see section 1 of the Act.
7. Provisions with a view to the adoption of the corresponding rules concerning receiving installations in harbours were laid down in Notice No. 379 of 27 July 1978, which has not yet come into force. See also Circular No. 128 of 27 July 1978.
8. More detailed provisions on the subject are set out in Notice No. 307 of 22 August 1968.
9. Section 3 of the Shipping Supervisory Act in conjunction with Royal Decree No. 163 of 19 March 1951 and Royal Decree No. 394 of 17 November 1952, both as subsequently amended.
10. Section 24 and 25 of the Act.
11. Section 26 of the Act. After 1 July 1980, see Ships' Safety Act, sections 11–13.
12. As from 1 July 1980, these provisions are replaced by more comprehensive rules concerning the right to issue orders and decide on detainment in section 15 and 16 of the Ships' Safety Act.
13. For further details, see B. Gomard in *U* 1956 B 102 ff. and 209 ff., and Kjeld Rosenmeyer, *Søret*, 1975, p. 226 ff.
14. Notice No. 162 of 2 May 1975.
15. Notice No. 449 of 29 August 1978; see also Notice No. 450 of the same date.
16. Ministry of Justice letter of 18 March 1980, printed in supplementary report of 19 March 1980 of the Danish Parliament's Environment Committee, p. 4 ff.
17. As regards how the previous regulations on limitation of liability, which were based on the 1957 convention concerning the limitation of shipowners' liability, relate to the current provisions concerning liability for damage due to oil pollution, see Allan Philip, in *U* 1971 B 153–55.
18. See for further details, Notice No. 163 of 2 May 1975.
19. See schedule to Chapter 5 of the Environmental Protection Act as amended by Notice No. 290 of 28 June 1978, subsection C, 1, 6 and 9, also subsection H, 2.
20. For the connection between the harbour regulations and the 1976 Act concerning commercial ports, see 2.3.2 with note 81.
21. Act No. 98 of 12 March 1980.

22. In this connection, see *Nyt fra miljøstyrelsen* No. 3/1976, North-Sea Oil and the Marine Environment.
23. See also in this connection, the Continental Shelf Act, section 2, subsection 1.
24. Notice No. 421 of 24 August 1976.
25. See the Continental Shelf Act, section 3.
26. See 5.1.3 above from which it is apparent that the Marine Environment Act applies to light as well as heavy mineral oils.
27. The information comes from the Minister of the Environment's comments on the Marine Environment Bill of 11 December 1979, see p. 19 in particular.

6
Discharge into the Sewerage System

Chapter 3 provides an account of the most important Danish legal rules concerning the discharge of polluting substances and liquids into fresh surface waters. This account focused on the evacuation of *waste-water* by which is meant all water that is discharged from dwellings, commercial undertakings, other buildings and paved areas (e.g. roads).[1] Owing to the structure of the legal material, the discharge of effluent into the sea from sources on land was included in Chapter 3, as the law generally places discharge into watercourses, lakes and the sea on the same footing.

Waste-water may, roughly speaking, be disposed of in three different ways: into the ground, with a view to its soaking away; into surface water (watercourses, lakes and the sea); and into a sewerage system, from which the waste-water collected is—after purification, if necessary—discharged into surface waters. In the case of both direct and indirect discharge to surface waters 'waste-water installations' are used, which the legislation interprets as both open and closed conduits and other installations serving to remove and process waste-water in conjunction with discharge into watercourses, lakes or the sea.[2] Hence waste-water installations include, on the one hand, installations for individual outfalls and, on the other, installations for outfalls which are common to several parties and, besides, purification installations. Whereas the legislation defines the terms waste-water and waste-water installation, there is no legal definition of 'sewer' or '*sewerage system*'. By these expressions I mean installations serving for joint disposal from a fairly large number of outfalls. It may be difficult to set an exact limit on the size of installation but, as a guide, it is reasonable to work on the assumption that installations receiving effluent from more than 30 person equivalents (approximately 10 households) are involved.[3] Sewerage systems in this sense will predominantly be 'public' which means according to the legislation that one or more local authorities is responsible for operating and/or maintaining the installation.[4]

The discharge of waste-water into the ground and special outflows to surface waters are covered by the account in Chapter 3. In this Chapter the question of discharging waste-water into the sewerage system will be dealt with. It should be added, however, that the regulations concerning disposal via the sewerage system and other forms of discharge, direct or indirect, into surface waters are so closely connected that some problems relating to discharge into what is here conceived as the sewerage system are discussed in Chapter 3.

6.1 BANS AND PERMITS

Section 17 of the Environmental Protection Act must be taken as the point of departure for a description of the system of regulation. Under this clause, it is forbidden to discharge into watercourses, lakes and the sea substances that may contaminate the water. This ban may nevertheless be set aside by special permits, in particular for the discharge, etc. of waste-water, see section 18 of the Act in this connection. These regulations are assumed to mean that both direct discharge into surface waters and discharge via joint waste-water installations require permission from an administrative authority. This means that discharge into the sewerage system also presupposes concrete approval.

The permit scheme and its provisions concerning which public authority (county council or municipal council) can grant permission for the discharge of waste-water is described in its entirety under 3.1.4.1.1, to which readers are referred.

It is apparent from the above explanations that it is of some importance for the division of authority that a *waste-water plan* should have been drawn up in the municipality concerned. Since the waste-water plan is, moreover, a deciding factor in the construction of the sewerage system and the establishment of purification plants, it is natural to describe the regulations concerning waste-water plans at this point.

Under section 21 of the Environmental Protection Act all municipalities are *obliged* to work out a waste-water plan. The municipal authorities were supposed to prepare drafts of such plans before 1 October 1976 and the final plan has to be approved by the relevant county council. By the middle of 1979 the county councils had approved approximately 150 waste-water plans in all.[5]

There are some outline provisions concerning the *contents* of the plans in section 21 of the Act which are amplified by the sewage Notice, the sewage Circular and some guidelines issued by the Environment Agency in October 1975.[6] The waste-water plan must, first and foremost, be an overall plan for the development of waste-water installations in the municipality. Besides information concerning the recipient situation and existing disposal of waste-water in the locality, the plan must contain in particular a plan for future sewerage and purification, detailing which waste-water installations are to be executed municipally and which by the landowners, and the dates for working out projects and establishing the installations, and also the financing arrangements.

A number of provisions have been drawn up concerning the *procedure* for preparing waste-water plans in sections 8–14 of the sewage Notice in particular. When the municipal authority has prepared a draft waste-water plan, this is to be submitted to the public for inspection. Moreover, at least one public meeting is to be held concerning the waste-water plan under the chairmanship of a county council representative. At least 3 weeks' notice of the meeting must be given and, among others, local organisations, some public authorities and specially affected landowners are separately notified of the meeting. Comments and objections may be submitted in writing before the obligatory public meeting and up to 3 weeks after it has taken place. On the basis thus provided the county council then decides whether to approve the plan.[7] In doing so, the county council may impose conditions for its approval, thereby altering the contents. The plan may be revised at any time subject to the procedure outlined being observed. This may be done on the initiative of the municipal authority and in accordance with instructions from the county council to prepare part-plans, amendments or supplements, see section 21, subsection 3, of the Act.

The waste-water plan has no direct *legal effects* for citizens in the sense that the latter can invoke a right to obtain discharge permits within the framework of the plan.[8] On the other hand, landowners have to set up their waste-water installations in accordance with the provisions of the plan concerning private installations, including the clauses relating to time limits, see section 23 of the Act. The plan, moreover, affects citizens indirectly in that the authorities are bound by the plan. They are, among other things, normally entitled to require connection to the sewerage system and the municipal authority can, when the waste-water plan has been approved, lay down rules for payment and demand contributions towards joint waste-water installations in the municipality under section 27 of the Act. If sewers are laid in an area, previous permits for discharge into soakaway installations and cisterns could be, and usually are, withdrawn, see section 11, subsection 3 of the Act.[9] If a landowner has been granted permission for a separate waste-water disposal system, connection with the sewerage system may be insisted on under section 25 if the existing installation 'does not function properly from the environmental viewpoint'.[10]

It should be added that the *environmental protection regulations* in Chapter 6 contain rules concerning *inter alia* the maintenance and cleaning of waste-water installations, so that they do not give rise to unhygienic conditions.

6.2 CONTROL OVER COMPOSITION AND QUANTITIES, ETC. FOR EFFLUENT DISCHARGED INTO THE SEWERAGE SYSTEM

6.2.1 Composition and quantities

In principle, decisions on these points are taken by the municipal authority and the county council in the individual discharge permit under section 18 of the Act. In this context the authorities are bound by the waste-water plan, including its provisions concerning the capacity of the waste-water installation.

As regards other material criteria for discharge permits, there are no binding regulations. On the other hand, in Guidelines No. 6/1974, the Environment Agency established certain recommended standards in this respect. Thus when industrial waste is discharged into the sewerage system, it should normally be stipulated that the pH must be in the range 6–10 and that the temperature must not be below 35 °C (subsection 4.2.10).

It is also stated that when granting permission for discharges from industrial plants, a reasonable safety margin should be allowed in the calculation in order, for instance, to keep a reserve capacity available, so that existing disposal permits may be extended and there is reasonable scope for new connections to be permitted. The requirements to be met by the individual discharger must be formulated in concentrations and quantities (g/24 hours, kg/per annum).

With a view to protecting purification plants, in particular, the guidelines state *inter alia* that many types of biological functions in the purification installation will be noticeably affected if, with uniform loading, concentrations in the inflow to the plant exceed the following levels (in mg/l):

Lead	1
Chromium (total)	2
Copper	1
Nickel	1
Silver	0·1
Zinc	2
Arsenic	1
Cyanide	0·1
Phenol	5
Mineral oils	10
Anionic detergents	10

These standards relating to the total inflow into the purification plant are significant by derivation for the requirements imposed on the individual discharger. In this context the guidelines state (subsection 4.2.13) that in many cases the upper limit for the proportions of the relevant concentrations contained in the individual discharge could appropriately be set at five times the levels that are valid for the inflow into the purification plant concerned.

6.2.2 Charges

Where there are several joint users of a waste-water installation (private or public) the question arises of payment for the building and operation of the installation. On this point the Environmental Protection Act stipulates that the costs in such cases must be borne and divided according to rules laid down by the municipal authority, see section 27. This rule applies to purification as well as other waste-water installations.

There are only a few provisions in the Act as to the bases on which the distribution of these costs can be decided by the municipal authority, although in a 1978 amendment to the Act not yet implemented the legal framework has been amplified to some extent.[11] The Minister of the Environment can, however, under section 28 of the Act lay down general rules for the municipal authorities' apportionment of costs. Such regulations have not so far been issued but in a 1974 circular the Ministry established some recommended standards.[12] The chief consideration in the distribution of costs is what might be called a *costing principle*. The Act expressly lays down in the new section 27, subsection 2, in regard to the installation as a whole that the landowners may not be assessed for a higher total contribution than the amount corresponding to the costs of execution and operation. In regard to the individual landowners too, however, the main rule is that the size of the contributions should be fixed according to criteria which lead to the costs imposed by the dischargers individually or as groups on the scheme as a whole being more or less covered.[13] This does not prevent the local authority from fixing the apportionment criteria with regard to the conditions for practical administration of the scheme. If the installation is in the general public interest, the municipality must, besides the costs which fall on the municipality's own properties, contribute a grant towards or allow deferred payment of the cost of building or operating the installation, see section 27, subsection 4, of the Act.[14]

It is apparent from the foregoing that charges cannot be laid down for waste-water disposal with the actual object of exercising environmental control; in this connection see 3.1.4.4 above concerning the use of charges as an instrument of environmental policy.

6.3 REQUIREMENTS FOR THE PURIFICATION OF WASTE-WATER

The Minister of the Environment is entitled to lay down general regulations concerning purification plants and the purification of waste-water under section 20 of the Environmental Protection Act but this has not been done hitherto. What purification plants are to be set up and to what extent waste-water is to be purified is determined in the first instance through the municipal drainage waste-water plan, see 6.2 above. It is also possible, with the county council's permission, to establish a new purification plant which is not covered by a waste-water plan—of particular importance where such a plan has not yet been approved.[15]

For the time being the Ministry of the Environment and the Environment Agency have adopted the general target for waste-water purification which was proposed by the Pollution Council in publication No. 15 of 1971. The idea of this was over a 10-year period, to aim to reach a level of purification which was designated 'purification level III' and would involve the following:

(a) Mechanical/biological purification and purification of nutritional salts (called chemical purification) for discharge into areas of still water (lakes and fjords), also watercourses debouching into these.

(b) Mechanical/biological purification for discharge into other watercourses.

(c) Mechanical purification for discharge into more open areas of sea.

It should be explained that this only represents a standard target and that the purification requirements must be substantially higher where very large quantities of waste are discharged into recipients of limited size.

DISCHARGE INTO THE SEWERAGE SYSTEM

With a view to fulfilling this target or creating others, the Ministry of the Environment and the Environment Agency have provisionally collected comprehensive details from the municipalities concerning *inter alia* existing and projected purification arrangements. The results of this study are published in the Environment Agency's publication No. 1/1974 *Waste-water purification 1972–1982*.

The study shows the following results in regard to the purification level in 1972:

Table 6.1 Purification level for the 4 types of recipient in 1972 plus % share of the 4 recipient types. 100% = 12 million person equivalents (pe)

1972	Unpurified (%)	Mechan. (%)	Mechan./biol. (%)	Mechan./chemical (%)	Mechan./biolog./chemical (%)	% share of recip. type
River–lake, lake, river–fjord, fjord	38	31	30	1	—	(45)
River–sea	16	42	42	—	—	(12)
Sea	66	17	17	—	—	(35)
Soakaway or unspecified	100	—	—	—	—	(8)
Total	50	25	25	0·5	—	(100)

Through implementation of the purification arrangements planned by the municipalities up to 1982, the purification level at that time will be:

Table 6.2 Purification level for the 4 types of recipient in 1982 plus % share of the 4 recipient types. 100% = 14·7 million person equivalents (pe)

1982	Unpurified (%)	Mechan. (%)	Mechan./biol. (%)	Mechan./chemical (%)	Mechan./biolog./chemical (%)	% share of recip. type
River–lake, lake, river–fjord, fjord	16	8	50	3	14	(45)
River–sea	3	19	75	1	2	(12)
Sea	6	30	63	1	—	(37)
Soakaway or unspecified	100	—	—	—	—	(6)
Total	16	17	50	2	6	(100)

These tables show that whereas in 1972 only 17% of waste-water was purified in accordance with the Pollution Council's target, in 1982 approximately half the waste-water will attain this standard of purification through a total investment of approximately 3·5 million Danish crowns in purification measures. This achievement in terms of purification will naturally mark a considerable advance but the result can scarcely be regarded as wholly satisfactory. It should nevertheless be noted in this connection, that it will, according to the study, be possible to attain purification level III generally in 1982 with the investment of about another 0·5 million crowns.

6.4 CONTROL OF DISCHARGE FROM SEWAGE AND PURIFICATION PLANTS

Drainage from these installations to surface waters takes place in accordance with the general provisions in Chapter 4 of the Sewage Notice. In this connection readers are referred to the account under 3.1.4 above, especially 3.1.4.1.1, 3.1.4.2 and 3.1.4.3.1.

6.5 ENFORCEMENT

In the case of public waste-water installations and major private joint waste-water installations for domestic effluent, among others, the county council and the municipal authority respectively must, under the Supervision Notice, normally inspect the recipient twice yearly and remove samples of the inflow to the outflow from purification installations.[16] See also 3.1.7.

6.6 CIVIL LAW LIABILITY IN DAMAGES

In this respect refer to 3.1.9.

Notes

1. See Sewage Notice, section 2, subsection 1.
2. See Sewage Notice, section 2, subsection 3.
3. See Sewage Notice, section 25, concerning the delimitation of the installations for which permits may be granted by the municipal council and the county council respectively. The term 'sewerage' is used in section 11, subsection 3, of the Environmental Protection Act. In this connection see too Environment Agency Guidelines No. 6/1974, p. 7.

4. See Sewage Notice, section 2, subsection 4.
5. See the Environment Agency report, *Miljøreformen*, November 1979, p. 35.
6. Ministry of the Environment Notice No. 174 of 29 March, 1974 and the same ministry's circular of 17 April, 1974.
7. As mentioned by Claus Tønnesen in *Dansk Miljøret*, vol. 3, 1977, p. 47 ff., the county council may not, in its decision concerning approval, deal with the more precise distribution of the financial charges between the landowners, as the county council would then be impingeing on the authority of the municipal council under section 27 of the Act.
8. See *Nyt fra miljøstyrelsen* No. 5/1976, p. 12 ff.
9. As the overriding rule, a landowner with such a method of disposal will be unable to avoid connection to the sewerage system, see for instance the Environmental Appeal Board's decision of 4 October 1976, *KFE* 1977, p. 46; see also *KFE* 1979, p. 64.
10. In Chapter 6.4 of the Environmental Protection Regulations a general obligation is imposed on owners of property adjacent to a new sewerage system to be connected to it. If the owner has hitherto legally discharged waste into surface water, however, it will scarcely be possible to enforce this obligation beyond the conditions laid down in section 25 of the Act.
11. Act No. 107 of 29 March 1978.
12. Ministry of the Environment circular No. 69 of 29 March 1974. Some importance as guidelines is also attached to the Minister's comments on the bill which led to the amendment of 29 March 1978 to the Act (bill of 11 October 1977, p. 6 ff).
13. See the statutory amendment mentioned in note 12, p. 6, and circular of 29 March 1974, point III.
14. For more details on the subject, see Claus Tønnesen, *op.cit.*, pp. 162–66, and Christian Hjorth-Andersen, in *Juristen og Økonomen*, 1978, pp. 455–60, the latter especially about the financing of public purification plants.
15. See Sewage Notice, sections 15, subsection 1, and 25, subsection 2.
16. Ministry of the Environment Notice No. 177 of 29 March 1974, section 5.

7
Waste Disposal on Land

Waste from consumption and production can arise in many different ways and may therefore occur in gaseous, liquid or solid form. Used in a broad sense, the concept of 'waste' pervades the preceding and subsequent Chapters alike. Nevertheless, it is inappropriate to deal with all forms of waste problems in just one section as this would involve considerable and unnecessary overlapping between the various sections. Therefore this section on waste will deal only with those questions that do not naturally slot into earlier and later Chapters.

The subject of this Chapter may be defined briefly using both positive and negative criteria. Roughly speaking, the following categories of waste will be discussed:

(1) Refuse, which is interpreted as ordinary household waste and similar waste from commercial undertakings.

(2) Special waste
    (a) Industrial and building waste, non-toxic packaging and other non-toxic waste.
    (b) Garden refuse, vegetable waste and the like.
    (c) Chemical waste, technical waste, sludge from industrial undertakings and other toxic wastes.
    (d) Oil waste.
    (e) Hazardous waste from hospitals.

(3) Sludge from waste-water purifying installations.[1]

These will be predominantly 'solid wastes' but liquid wastes and intermediate forms between liquid and solid are also included.

On the negative side, this Chapter will not be concerned with gaseous waste (smoke), waste water, refuse from ships and platforms, radioactive waste and the disposal of toxic substances. On these subjects, readers are referred to Chapters 2, 3, 4, 9 and 10, respectively.

This survey is based in principle on differentiating between ordinary waste (7.1–5 below) and special forms of waste (7.6 and 7.7 below). This division does not coincide with the classification set out above. According to Danish law, it is convenient to deal under one heading with (1), (2) (a) and (b) and also (3), whilst special wastes include mainly only (2) (c)–(e).

7.1 CONTROL OF THE SITING OF INSTALLATIONS FOR THE DEPOSIT AND TREATMENT OF WASTE

The siting of these installations is governed, on the one hand, by the general planning legislation and, on the other, by the Environmental Protection Act. In regard to the planning legislation, refer to the account under 2.1.1. It need only be added that the regional plans will include provisions concerning the future location of dumping sites and incineration plants and that these provisions are being drafted in the light of environmental quality planning under Chapter 9 of the Environmental Protection Act.[2]

Control under the Environmental Protection Act is based on Chapter 5 thereof which stipulates that installations for the storage and treatment of more precisely defined forms of waste are subject to prior approval by the environmental protection authorities—for further details see point G of the schedule and also D.10. This requirement regarding approval applies to practically all forms of installation for storing and treating waste. The precise formulation of the list was altered somewhat in conjunction with the 1978 amendment but it was, in the main, merely amplified by codifying practice under the original list. The decision on approval is in most cases taken by the county council.

To assist the county councils in deciding whether to approve dumping sites, the Environment Agency circulated its recommendations in Guidelines No. 1/1974 on controlled dumping sites. The approval scheme under the Environmental Protection Act is, moreover, dealt with in Section 2.1.2.[3]

7.2 REQUIREMENTS CONCERNING TREATMENT PRIOR TO DEPOSIT OF THE WASTE, WITH A VIEW TO RECYCLING

There are no current regulations imposing general requirements for the 'pretreatment' of waste. On the other hand, Chapter 4 of the *Recycling Act*,[4] which came into force on 1 January 1979, authorised the Environment Minister to issue certain regulations of this type, namely regulations concerning the collection of waste in the form of paper and drink packs with a view to recycling.[5]

This legislation was preceded by fairly comprehensive investigations under the direction of the Environment Ministry, especially into the possibilities of recycling paper. In March 1975 the Environment Minister made a general statement to the Folketing on recycling problems and opportunities which was the subject of a parliamentary debate.[6] This was followed in October 1975 by a longer report from a working party under the Environment Ministry on 'recycled paper' and the conclusions in this report significantly influenced the contents of Chapter 4 of the Recycling Act.[7]

The Recycling Act is intended, on the one hand, to prevent and combat pollution and, on the other, to restrict the consumption of new raw materials. With the continuous growth in volume of waste, the problems of pollution associated with waste disposal also mount. An obvious solution is to endeavour to reduce the amount of waste for disposal or to change its composition so that it is easier to treat or store. At the same time, society's consumption of raw materials, both mineral and vegetable, is so great as to jeopardise future raw material supplies in certain sectors, in the long term at any rate. A common partial solution to both of these problems is to ensure that substantial portions of the waste are re-used.

The first step in this direction was taken with the Recycling Act. This covers solely (1) paper, cardboard and board materials and products made thereof and (2) packs used for drinks and consumer milk products together with raw materials, additives and auxiliary substances which are used in the production of the materials and products indicated under (1) and (2).

Chapter 4 of the Act empowered the Environment Minister to lay down regulations making the local authorities responsible for the collection of recyclable materials and products within the scope of the Act. The Act provides for the more detailed organisation of arrangements in the individual municipality to be left to the local authority's discretion. Thus the municipal council may, under section 11, subsection 2, decide to entrust collection wholly or partly to appropriate enterprises or organisations but the final responsibility for collection lies with the local authority. By 'appropriate enterprises or organisations' are understood *inter alia* industrial recycling enterprises and scouting organisations. The municipal council may, moreover, issue a by-law on the collection scheme, stipulating therein *inter alia* that one or more types of waste shall be kept separate from other waste, cf. section 11, subsection 3. If a compulsory collection scheme is set up, the Minister is bound to stipulate simultaneously that the local authorities must, on certain conditions, exempt commercial undertakings, for instance those providing proof that re-usable materials and products are sold for recycling purposes.

The Act is, in general, based on the assumption that regulations will only be implemented in the sector if they are appropriate from the socio-economic viewpoint, cf. too in this connection the definition of aims in section 2, subsection 1.

The Minister's comments on section 11 of the Bill, corresponding to section 11 of the definitive Act, state that the local authorities 'ought' to organise the collection scheme in such a way that the costs do not exceed the income from the sale of recycled products and the savings resulting from the reduced amounts of waste.[8] As the possibility of the scheme involving net costs for the municipality cannot be ruled out, section 11, subsection 5, of the Act empowers the local authority to fix charges to cover these.

So far the Environment Minister has not laid down regulations under section 11. This is partly because the prices for recycled products have been very low for a time as a result of the economic recession. A final point is that the Minister can only exercise his powers under the Act to lay down regulations after consulting *inter alia* the municipal organisations and the national business and consumer organisations most closely concerned, cf. Chapter 3.

7.3 CONTROL OF THE METHOD OF DISPOSAL

The problems of disposal raise three questions in particular:

(1) Are consumers and undertakings obliged to participate in a joint removal scheme?

(2) How is the waste to be treated or stored after removal?

(3) What regulations apply to the establishment and operation of installations for the storage and treatment of waste?

7.3.1 Compulsory removal schemes

When dealing with this subject, a distinction must be made in what was described above under one heading as 'ordinary waste' between refuse (household waste), animal waste and other ordinary waste. The general provisions of Chapter 4 of the Environmental Protection Regulations apply to all three categories,[9] but these are supplemented by special legislation on some points.

As regards *refuse* (household waste), by which is meant waste originating naturally from households and similar waste from enterprises, etc. the municipal council can always decide that this waste should be removed under a common scheme in which all householders are obliged to participate and for which they must pay. As regards built up areas with more than 1000 inhabitants (permanent or temporary) the local council must arrange for communal refuse removal. When a compulsory communal refuse scheme has been introduced in the municipality, the local council may issue more detailed regulations in a by-law concerning *inter alia* waste containers, their use and emptying and also the transport of the refuse. By far the majority of Danish households, etc. participate in joint disposal schemes for household waste and it is estimated that by 1985 the number will attain 95%.[10]

With *animal waste*, the basic assumption is that it is removed by the enterprise itself for burning, to a destructor plant, or for processing into feeding stuff in accordance with the stipulations of the Environmental Protection Regulations or special legislation concerning meat and fish. If this waste cannot be properly disposed of by private arrangement, the local council may set up a compulsory joint scheme for the purpose, however. The provisions of the regulations concerning containers and vehicles for the transport of animal waste apply and the local council may lay down supplementary regulations. As regards the disposal of certain forms of meat and fish waste, however, more detailed regulations are issued by the Minister of Agriculture or Fisheries.

For *other forms of waste* (e.g. various types of building waste and non-hazardous industrial waste) there are no detailed regulations, but Chapter 4.4 makes provision for the local council to exercise control. If the storage, transport or disposal of such waste results in unhygienic conditions or seriously interferes with the surroundings, the municipal council can issue orders relating to special means of storage and disposal. If the waste cannot be satisfactorily disposed of by the private sector, the local council also has the right to introduce a compulsory communal scheme.

7.3.2 Treatment or storage of waste after removal

In principle, the local council alone decides on the method of disposal, primarily whether the waste is to be deposited (composted) at a dumping site or burnt in an incineration plant. This principle is modified in regard to animal waste, cf. 7.3.1 above. The basic rule may, moreover, be modified by means of regulations issued in pursuance of section 11 of the Recycling Act, see 7.2 above.

In the early 1970s disposal by means of incineration plants was considered the most efficient and hygienic method of disposing of refuse.[11] During that decade the proportion of waste incinerated rose, so that in 1977 1·2 million tonnes out of 2·15 million tonnes of waste suitable for treatment was incinerated whilst 0·14 million tonnes was composted. Today incineration is not considered unquestionably the best method and it is anticipated that in 1985, of a total amount of waste suitable for treatment of 2·7 million tonnes, 1·64 million tonnes will be incinerated and 0·285 million tonnes composted.[12] As regards dumping sites, the number has been falling sharply over the last few years and in 1985 there are expected to be only about 60 controlled dumping sites in the whole country.[13]

7.3.3 Establishment and operation of installations for the deposit or treatment of waste

In this respect, too, it is initially up to the local council to reach decisions on the *establishment and operation of dumping sites and other installations for waste disposal.*

This includes issuing directives about the pretreatment of waste by grinding or compression, about covering the waste, and about supervising the site. Nevertheless, the local council does not have a free hand in regard to such regulations. On the one hand, in relation to approval of dumping sites, etc. special conditions may be laid down concerning, for instance, pretreatment or supervision and, on the other, the Environment Agency's guidelines mentioned above in 7.1 have a certain regulatory effect. Thus the guidelines contain quite strong recommendations on several points concerning the operation of dumping sites. For instance, for sites that receive untreated waste it is stated that the waste should be spread in a layer at most 2 m thick and compressed by bulldozer, compactor or the like and be covered at the end of each working day with at least 15 cm of mineral earth. The guidelines also specify that the products of combustion from incineration plants must, after deposition on dumps, be covered by watertight means so that surface water is, as far as possible, prevented from coming into contact with the combustion products. If such directives are not complied with, proceedings may be instituted in this connection before the Environment Agency upon complaint by, for example, the medical officer, cf. 1.7.3 above.

7.4 RESTORATION OF THE GROUND AFTER TIPPING

There are no general regulations on this subject. On the other hand, requirements may be specified as to soil treatment and planting after the dumping site has been filled in at the time of its approval in pursuance of the Environmental Protection Act or the Nature Conservation Act. In this connection it is observed that the above-mentioned Environment Agency guidelines, when discussing points that should be clarified before approval is given, state that all the main features of the aspect of the area and its utilisation after it has ceased to be used as a dump must be established.

7.5 ENFORCEMENT

Supervision of compliance with the provisions of the Environmental Protection Act, the Environmental Protection Regulations and the waste by-laws (issued under the regulations) is predominantly exercised by the local authority. No more specific provisions concerning supervision exist but they may be issued by the Environment Minister under section 57 of the Environmental Protection Act. For dumps operated by the municipalities, which is the normal practice, supervision is exercised by the county council, cf. section 50 of the Act. It is noted, moreover, that 'neighbourhood supervision' plays an important role in practice for nuisances due to dumps other than water pollution, as has been proved by a large number of complaints.

As regards supervising compliance with the regulations presumably in the offing to implement the Recycling Act, section 14 thereof provides that the Environment Minister shall lay down more detailed directives.

Infringement of any of the above-mentioned regulations constitutes a punishable office; in the sole case of waste by-laws issued by municipal authorities punishment is by fine only. Prosecutions are brought in accordance with the rules of ordinary legal proceedings. As far as is known, there have been no prosecutions in this sector to date.

For decisions under Chapter 5 of the Environmental Protection Act (e.g. concerning approval of or orders relating to a dump), an appeal lies to the Environment Agency and thereafter to the Environmental Appeal Board, cf. 1.7.3 above. If the decision has been reached under the Environmental Protection Regulations or the waste by-laws, the appeal lies to the Environment Agency and exceptionally thereafter to the Environment Ministry.[14] The same applies to forthcoming regulations concerning waste collection under the Recycling Act.[15]

7.6 CONTROL OF SPECIAL FORMS OF WASTE

It is difficult to draw the line between 'ordinary' and 'special' waste and the demarcation is to some extent arbitrary. Among wastes that may be termed 'special' it is furthermore hard to determine how far to go in specifying these, particularly in a general survey. Here we are chiefly concerned with *whether* the waste involves special environmental hazards and *whether* this has resulted in specific mandatory provisions or recommendations. Matters relating to oil and chemical waste take pride of place (7.6.1 and 2) but under 7.6.3, dealing with other forms of special waste, there are a few comments on hospital waste, fly ash and used-car dumps. In regard to the two latter forms of waste, the borderline with what may be described as ordinary waste is close.

7.6.1 Waste oil

Under Act No. 178 of 24 May 1972 regulations were issued concerning *inter alia* the disposal of oil and chemical wastes. The Act is primarily an enabling act which authorises the Environment Minister to lay down more detailed regulations concerning compulsory participation in *inter alia* a reporting and delivery scheme, and contains no directly operative provisions.

In regard to oil waste, the powers were utilised by means of Notice No. 455 of 17 October 1972. On the basis of experience acquired and investigations undertaken and in fulfilment of the EEC Directive of 16 June 1975 on the disposal of waste oil, the standards were substantially tightened up in many respects with the issue of Notice No. 410 of 27 July 1977, which contains the current regulations.[16]

The 1977 Notice is more comprehensive that its predecessor as the concept of waste oil is wider than in the 1972 Notice. The current Notice interprets 'waste oil' as all oleaginous products which are no longer intended for use for their original purpose in their present state, cf. section 1, subsection 3. The amplification consists in the inclusion, on the one hand, of saleable oil products, provided they are no longer destined for their original purpose, and, on the other, of waste from synthetic oil products as well.[17]

Undertakings where waste oil in this sense arises must in accordance with Chapter 4, *notify* the local authority of this unless the quantity is less than 150 litres per annum. Under Chapter 5, moreover, these undertakings must *deliver* their waste oil to a place designated by the local council, unless the waste is collected by arrangement with the local authority. The delivery obligation is not subject to a minimum quantity, as in the case of notification, but the local council may exempt an enterprise from the duty to deliver upon proof that the oil waste is acceptably transported and disposed of on the enterprise's own initiative. Such exemption may be granted, for instance, if the enterprise delivers its waste oil direct to a reprocessing plant.[18] In conjunction with the delivery obligation, the municipalities were directed to set up local reception arrangements before 1 September 1977.[19]

Besides the control inherent in the notification and delivery arrangements, the Notice grants powers of intervention to the local councils. The latter may, in the first instance, lay down more detailed rules in the form of by-laws to ensure that waste oil is properly stored and transported, cf. section 2, subsection 2. To the same end, the local council may issue specific orders. If pollution is caused during storage, transport or disposal of waste oil, the local council can, finally, issue orders concerning the elimination of the pollution, cf. section 3, subsection 2. It must be assumed that these powers on the part of the local council may be used notwithstanding the legal protection afforded to an enterprise approved under Chapter 5 of the Environmental Protection Act by section 44, subsection 4, of that Act.[20]

7.6.2 Chemical waste

The powers granted to the Environment Minister under the 1972 Act were first utilised in Notice No. 121 of 17 March 1976. This contains the current regulations on chemical waste with a minor amendment in 1980.[21]

The scope of the Chemical Waste Notice is, according to section 1, indicated primarily by a schedule which forms an annex to the Notice. The following somewhat vague provision is appended to the schedule:

> ... together with such other kinds of chemical waste as have similar characteristics, e.g. caustic, toxic or inflammable.[22]

In terms of content, the regulations under the Chemical Waste Notice correspond in all essentials to the arrangements for waste oil. Thus in the chemical waste sector too there is a duty to notify and deliver, the municipalities are bound to establish local reception installations and the local councils have powers of intervention, both general and specific. The differences which may be mentioned are that compulsory notification of chemical waste applies irrespective of quantity and that enterprises wishing to import chemical waste from abroad must notify the local council thereof at least one month before importation. There is a further difference arising from a special scheme concerning the 'chemical waste card'. Under section 2, the Environment Agency has to prepare chemical waste cards for all the types of waste listed in the annex to the Notice and, under section 5, subsection 2, such a card has to be issued to the enterprise and accompany the waste in transit. The cards contain guidelines on, for example, packing, marking and measures to be taken in the event of accident.[23]

Virtually all chemical waste delivered to the municipalities is finally processed together with waste oil by one undertaking, Kommunekemi A/S, which is predominantly owned by the primary municipalities' organisation, Kummunernes Landsforening ('National Association of Municipalities').

7.6.3 Other special forms of waste

Hospital waste Hospitals are the source of ordinary refuse, on the one hand, and special hospital waste, on the other. The first can, of course, be disposed of in accordance with the general regulations covering refuse, cf. 7.1–7.3 above. However, some hospital waste involves special problems, for instance waste from patients with infections, pointed and sharp objects such as syringes, bandages and left-over food from wards. Insofar as this counts as chemical waste, it is dealt with under the regulations mentioned in 7.6.2. Other special hospital waste is governed chiefly by the very vague provisions concerning 'other waste' in Chapter 4.4 of the Environmental Protection Regulations, cf. 7.3.1 above. However, in its Guidelines No. 1/1976 the Environment Agency laid down more precise recommendations for the disposal of special hospital waste.

Fly ash The disposal of fly also takes place under the general regulations referred to under 7.1–7.3 above. The storage of fly ash must be approved under Chapter 5 of the Environmental Protection Act, cf. point G.1 of the list, and the method of disposal is covered by Chapter 4.4 of the Environmental Protection Regulations. In addition, the enterprises producing fly ash—typically power stations and incineration plants—are normally themselves subject to approval under Chapter 5 of the Environmental Protection Act and conditions concerning disposal of the fly ash are normally linked with the approval. In *Nyt fra miljøstyrelsen* No. 3/1979, the Environment Agency provided information on the existing regulations concerning the deposition of fly ash and also set out a number of guidelines including some on the re-establishment of areas where fly ash has been deposited.

Vehicle scrap Sites for storing vehicles unfit for use (used-car dumps) have to be approved under Chapter 5 of the Environmental Protection Act and the collection provisions of Chapter 4.4 of the Environmental Protection Regulations also apply to this waste. Among provisions of special importance for used-car dumps section 44 of the Nature Conservation Act may be mentioned.[24] In accordance with this regulation, the county council—and in the metropolitan area, the metropolitan council—may order the erection of fences, planting or other remedial measures if a storage site constitutes an appreciable nuisance or looks very unsightly in relation to the surroundings.[25]

7.7 CONTROL OF SPECIAL CATEGORIES OF PRODUCTS

An obvious method of restricting the amount of waste is by preventive control, e.g. by banning certain types of non-returnable packaging and the use of particular substances and materials which may pose special waste disposal problems.

Legislation of this type was passed with Act No. 293 of 9 June 1971 relating to containers for beer and soft drinks (the *Can Act*). Under this

Act the Environment Minister could take steps to prevent the retail sale of beer and minerals in non-returnable packaging in Denmark. This might include the adoption of regulations to prohibit or make compulsory the use of certain types of packaging and requiring that the containers must comply with specific requirements as to their nature and weight. In the first few years after the Act was passed it was used only as the background to an 'agreement' between the Environment Agency and the breweries to restrict the total sales of beer, etc. in cans. By means of Notice No. 136 of 5 April 1977, the authority contained in the Act was transformed into mandatory regulations. With effect from 1 June 1977, the retail sale of a number of carbonated beverages—in particular beer and soda water—in non-returnable packaging was thereby prohibited. The Environment Agency may grant exemption from the ban in special cases and it is, in fact, granted on a limited scale.

The Can Act was repealed when the 1978 *Recycling Act* took effect, cf. 7.2. The provisions of this Act cover the same sector as the Can Act, cf. section 8 of the Recycling Act, plus a substantially wider field. Among other things, under section 6 of the Recycling Act the Environment Minister may lay down regulations stipulating that specific raw materials, additives or auxiliary substances which prevent the re-use of materials or products must not be present in materials or products that are sold or used in Denmark.[26] In addition, the Minister may, in pursuance of section 7, prescribe that paper, cardboard and board materials and products made thereof must contain specific proportions of previously used and re-usable materials and products. Finally, it should be mentioned that the Minister can, pursuant to section 9, lay down regulations concerning deposit arrangements for particular types of packaging, including fixing the amount of the deposit. As yet, no notices have been issued under the Recycling Act and the 1977 Notice is expressly upheld by the Recycling Act.

7.8 INDIVIDUALS' RIGHTS

In this connection readers are referred mainly to the account under 2.1.8 and to further references in 1.7, but one or two special points will be highlighted here.

Whereas the Recycling Act contains a *confidentiality* rule corresponding to section 85 of the Environmental Protection Act as amended in 1978 and does not therefore restrict the public's *access to documents* under the Public Access to Documents Act, the confidentiality provision in section 1, subsection 3, of the Act on Oil and Chemical Waste is 'special' in relation to section 7 of the Public Access to Documents Act.[27] This means that the public do not have access to documentary information concerning the nature, amount and composition of oil and chemical wastes. This difference between arrangements under the three environmental acts does not appear to be completely justified.

The *rules of appeal* in the legislation relating to oil and chemical waste and in the Recycling Act largely resemble the arrangements under the Environmental Protection Act. It is noted that only in respect of what are considered the most important decisions for those involved does an appeal lie to the Environmental Appeal Board.[28]

Questions of *liability in damages* in conjunction with damage caused by buried chemical waste have cropped up in recent years because there have been several discoveries of chemical waste buried in the past of which the authorities and often the present landowners too were unaware. So far these cases have not resulted in a judgment being given in this sector. In a 1978 article Bernhard Gomard took stock of the legal situation concerning liability in damages for environmental damage caused by chemical waste.[29] His main conclusion was that the owner of the property or installation in question was responsible in accordance with the general *culpa* rule, i.e. only if his actions had been unjustifiable or if he had managed his property or the installation unjustifiably. This *culpa* liability also covers cases where a dangerous state existed or developed on a property or in an installation and was known or should have been known to the owner and which could then have been averted by a repair, safety measures or the like, even if the owner was not to blame for the hazard present. Gomard assumes, furthermore, that the owner is bound to pay the costs of averting damage due to pollution, for example, by defraying the public authorities' expenditure on the removal of buried chemical waste and soil thereby contaminated. In regard to Gomard's views on liability for the authorities' expenditure on tracing buried waste, readers are referred to the account above under 3.1.9.

Notes

1. Cf. *Nyt fra miljøstyrelsen* No. 10/1974, p. 2.
2. Cf. Environment Ministry circular of 22 February 1977.
3. Cf. for further details of administrative practice concerning approval of dumping sites and the like, Jens Christensen, *Særlig forurenende virksomhed* ('Highly polluting industry'), 1980, pp. 185–231.
4. Act No. 297 of 8 June 1978.
5. Other parts of this Act are discussed under 7.7.
6. The Minister's statement and the subsequent parliamentary debate are reproduced in *Nyt fra miljøstyrelsen No. 5/1975. In this connection, see also Nyt fra miljøstyrelsen* No. 8/1975 which reports some ministerial replies to questions from the parliamentary Environmental Committee concerning the recycling of paper.

7. Issued by the Environment Ministry in the publication *Returpapir* ('Recycled paper'), 1975. See also *Nyt fra miljøstyrelsen* No. 1/1978, which reports on the possible uses of paper produced from recycled materials.
8. Cf. *Folketingstidende* 1977–78, appendix A, col. 3371.
9. Environment Ministry Notice No. 170 of 29 March 1970 with subsequent amendments.
10. Cf. Environment Agency publication *Miljøreformen*, 1979, p. 187.
11. Cf. Environment Agency publication *Miljøreformen*, 1979, p. 39.
12. *Op.cit.*, p. 187.
13. *Op.cit.*, p. 188.
14. The appeal scheme also covers the municipal council's decisions concerning exemption from participation in the compulsory refuse arrangements. For further details see Claus Haagen Jensen, *op.cit.*, p. 135; also *FOB* 1977, p. 379, where such a decision was appealed against up to the level of the Environment Ministry which, in the ombudsman's opinion, wrongly neglected to discuss the merits of the case.
15. Cf. sections 12 and 13 of this Act. Concerning appeals under the Recycling Act otherwise, see 7.8 and note 28.
16. For further general information on the current system, see Environment Ministry circular of 1 September 1977.
17. Regarding waste from synthetic oil products, demarcation problems may arise in relation to the Chemical Waste Notice; for further details see subsection 2 of the circular mentioned in note 15.
18. In regard to exemptions and information for dealing with applications for these, see *Nyt fra miljøstyrelsen* No. 2/1980.
19. See subsection 3 of the circular mentioned in note 15.
20. As far as the Chemical Waste Notice is concerned, this stems directly from section 14. No such provision exists in the Waste Oil Notice but, by virtue of the general rules of interpretation, the outcome must be deemed to be the same.
21. Notice No. 323 of 3 July 1980. For further information on chemical waste, see Environment Ministry circular of 14 October 1976.
22. Some points of doubt concerning the scope of the annex are referred to in subsection 2 of the circular mentioned in note 20.
23. Some guidance for municipal councils and enterprises is to be found in *Nyt fra miljøstyrelsen* No. 3/1980, which contains a list of sectors of industry indicating the typical forms of chemical waste that arise from various kinds of operations.
24. Statutory Notice No. 435 of 1 September 1978.
25. It is explained in *Nyt fra miljøstyrelsen* No. 5/1975, p. 9, that there are in Denmark approx. 600 dumping sites for cars, including approx. 300 with between 3 and 25 cars only.
26. Readers are reminded in this connection that the Act covers only paper, cardboard and board materials, also products made thereof, packs for beverages and consumer milk products and raw materials, additives and auxiliary substances used in the production of these materials and products.
27. Cf. in this connection Oluf Jørgensen, in *Grænser for deltagelse* ('Limits to participation'), 1980, p. 157, and 1.6.3 above.
28. See in this connection section 10, subsection 4, of the Waste Oil Notice, section 11, subsection 4, of the Chemical Waste Notice and section 13 of the Recycling Act. A few special rules concerning the function of the Environmental Appeal Board in recycling cases are given in Notice No. 18 of 11 January 1980.
29. *U* 1978 B pp. 53–71, especially p. 66 ff.

8
Noise and Vibration

8.1 STATIONARY SOURCES

Here questions of noise and vibration from fixed installations will primarily be dealt with. It is nevertheless reasonable to include under this heading transportable equipment which is normally used in conjunction with a specific property.

8.1.1 Control over the siting of installations which produce noise and vibration

The *general planning legislation* provides important means of influencing the location of new installations of this type (see especially 2.1.1 above). It is largely possible to ensure thereby that installations involving noise and vibration are so sited as to avoid nuisance and hence to eliminate such pollution since, unlike other types of pollution, it is not in their nature to 'accumulate' in the environment. It is noted here that, as an element in environmental quality planning under sections 61 and 62 of the Environmental Protection Act, the noise load *inter alia* is to be charted within each individual county and that, on this basis, targets are to be fixed for future environmental quality in regard to noise.[1] These environmental quality plans are incorporated in the regional planning. The existing draft regional plans indicate that the plans will contain limits for noise from, for instance, works, roads, railways and airfields, and that these limits will by and large correspond with the standards in the Environment Agency's guidelines and other generally accepted standards.[2]

The control apparatus for this purpose is extended by the approval regulations in Chapter 5 of the Environmental Protection Act; see 2.1.2

above for further details. Among the 'heavily polluting enterprises', etc. listed in the schedule drawn up under the Act which may not be established and put into operation until approval has been granted by the local council (county council or metropolitan council), there are some enterprises, installations and equipment for which the approval requirement is based primarily on their noisiness. Examples are shipbuilding yards, cement-casting works, saw-mills, motor racetracks, airports and shooting ranges. Among installations not covered by the approval scheme, there is special reason to point out railways and roads, but for further details on that subject see below.

The approval decision is taken, in cases based on noise and vibration, after an assessment. This takes account primarily of the polluting effect of the installation, after the execution of remedial measures if applicable, but the social utility, etc. of the enterprise also enters into the assessment, cf. section 1 subsection 3 of the Act. However, the Environment Agency's guidelines on noise, which are discussed more fully under 8.1.3 below, have resulted in a substantial standardisation of the basis of decision in these cases.

A few special comments are called for concerning roads, railways and airports. In regard to *roads*, it must be pointed out initially that road traffic is responsible for the greatest and most widespread noise nuisance. Almost half of the population is exposed to outdoor noise levels of 55 dB (A) or over and about 20% to outdoor noise levels of over 65 dB (A).[3] According to a survey by the *Socialforskningsinstituttet* ('Social Research Institute') about 40% of the population feels disturbed by noise and road noise is plainly the most frequent cause of nuisance.[4]

As mentioned, roads are not covered by the approval scheme under the Environmental Protection Act or by any other of the regulating provisions of that Act. Under section 10 of the Act the Environment Minister may, after consulting the Transport Minister, lay down regulations to the effect that projects for 'major road schemes and for railways' must be submitted to him before work is commenced but such regulations have not been issued so far. On the other hand, road schemes are controlled, apart from the general planning legislation, by separate road legislation. Under the Public Highways Act (termed the 'Roads Act' below),[5] public roads may be build only in accordance with a road plan drawn up by local councils and county councils and approved by the Transport Minister. The siting of trunk roads, including those that are being developed as motorways, is determined directly by law, however. This enables the noise effect of the road scheme to be taken into account in relation to the present or future use of the surrounding area but traditionally the location of road schemes has been predominantly determined by other considerations.

STATIONARY SOURCES

As a basis for the weight to be given to road noise in the planning of roads, both under the general planning legislation and the special road legislation, the Environment Agency in *Guidelines* No. 2/1974[5] set out the following figures for sensitivity to noise as equivalent continuous sound levels:

1	2	3
Type of area/Urban function/ Building development	Satisfactory environment with equivalent continuous sound levels at or lower than those shown below	Unsatisfactory environment with equivalent continuous sound levels higher than those shown below
Summer cottage areas, recreation areas outside urban areas	40 dB (A)	50 dB (A)
Residential areas, hospitals, recreation areas in urban areas	45 dB (A)	55 dB (A)
Hotels, churches, theatres, offices	50 dB (A)	60 dB (A)
Enterprises with low internal noise levels, small businesses	55 dB (A)	65 dB (A)
Large stores, supermarkets	65 dB (A)	75 dB (A)
Enterprises with high internal noise levels	70 dB (A)	80 dB (A)

The figures indicated for noise are outdoor free-field noise levels determined in accordance with ISO Recommendation 1996. The text of the guidelines states *inter alia* that in regard to the internal road systems of major urban areas a substantial part of the 10 dB (A) which constitutes the difference between the noise levels in columns 2 and 3 must in practice be achieved by measures at the detailed stage of planning, i.e. in conjunction with building and incidental plans, especially screening measures, etc.

A committee set up by the Transport Minister suggested in a 1978 report on road noise[6] that *inter alia* the regulations of both the Environmental Protection Act and the Roads Act should be extended with a view to preventing noise nuisance. The incorporation of a new subsection in section 10 of the Environmental Protection Act was proposed, authorising the Environment Minister to establish binding noise limits for new major road schemes and, as the normal maximum limit for roads carrying a traffic volume of more than 1500–2000 vehicles per day, the committee recommended 55 dB (A).[7] The committee wanted the Roads Act to include provisions obliging the highway authorities to ensure compliance with the abovementioned noise limits, either through the

NOISE AND VIBRATION

siting of the road, screening or facade insulation.[8] Should the cost of noise-insulating measures be out of all proportion to the value of the property affected by noise, the highway authority would offer to take it over against compensation. Legislation to this effect was in the course of preparation but was subsequently abandoned for the time being,[9] as far as is known chiefly because of the municipalities' objection to being put to further expense. It must also be acknowledged that if the noise limits recommended by the committee are incorporated in the regional plans—as seems very likely—the committee's proposals concerning compulsory noise limits are then superfluous but the proposed regulations concerning the duty to take precautions for noise abatement still hold.

Like roads, *railways* are not covered by the Environmental Protection Act. There is, however, no general legislation governing the planning of new railway routes, as there is with roads, since such issues are customarily decided directly by special installation acts. On the other hand, railways are, in principle, covered by the general planning legislation and the environmental quality planning associated therewith. In regard to the assessment of train noise, there is no generally recommended standard, as for roads, since apparently no suitable method of calculating train noise or criteria for acceptable train noise loads are known as yet.[10]

Airports and other airfields must, unlike road schemes and railways, be approved anyway under Chapter 5 of the Environmental Protection Act, see point 1, 2, of the schedule, and the approval decision is reached in the first instance by the county council concerned. In addition, permission to establish and operate airfields is required to some extent in pursuance of the Aviation Act.[11] Airfields which are open for use by the public must always receive permission from the Transport Minister and this Minister may similarly decide about other airfields. Permission may only be granted if deemed consistent with 'general considerations', which include the polluting effect of the enterprise, and conditions may be attached to the permission. The siting of major airports is generally determined by special legislation. Furthermore, airfields, like roads and railways, are in principle covered by the general planning legislation and the associated environmental quality planning.

According to the Environment Agency's Guidelines No. 2/1974 concerning 'environmental considerations and planning', the location of new airfields should occur either in conjunction with agricultural or forestry areas not sensitive to noise or, if necessary, closely linked with other noisy transport installations and industry. As a guide for considering the environmental aspects of the location of future airfields, the Environment Agency sets out in the guidelines the following noise levels as satisfactory and unsatisfactory environments respectively for various types of area, etc.:

1	2	3
Type of area/Urban function/ Building development	Satisfactory environment with noise levels at or lower than those shown below	Unsatisfactory environment with noise levels higher than those shown below
Summer cottage areas, recreation areas outside urban areas	80 PNdB	90 PNdB
Residential areas, hospitals, recreation areas in urban areas	85 PNdB	95 PNdB
Hotels, churches, theatres, offices, enterprises with low internal noise levels, small businesses	90 PNdB	100 PNdB
Large stores, supermarkets	95 PNdB	105 PNdB
Enterprises with high internal noise levels	100 PNdB	110 PNdB

The table is based on the maximum noise level at ground level expressed in PNdB and measured by a modified American method described more fully in the Pollution Board's publication No. 25 concerning aircraft noise, annex 5.5. The comments on the table in the guidelines mention *inter alia* that around major airports serving regional centres where a central position in relation to the mass of population carries considerable weight, levels higher than the figures in column 2 will have to be accepted but that only exceptionally should the levels in column 3 be allowed to be exceeded.[12]

8.1.2 Control over the establishment and use of noisy installations and equipment

Binding *general regulations* on these matters may be laid down by the Environment Minister under section 6 of the Environmental Protection Act. Hitherto these powers have only been exercised in a 1980 Notice to issue regulations concerning the noise from motor lawnmowers.[13] This stipulated, in particular, that as from 1 October 1980 motor lawnmowers that are offered for sale must not exceed more specific noise limits which are tightened up with effect from 1 October 1983. It will also be possible to apply section 6 of the Act to many other types of installation, machines and tools, e.g. compressed-air tools and compressors.

NOISE AND VIBRATION

Furthermore, it should be mentioned that Chapter 9 of the building regulations of January 1977 (issued pursuant to the Building Act) contains construction provisions designed to limit the noise that occurs in the building concerned. It is stipulated *inter alia* that communal technical installations (for instance, lifts and heating systems) shall be executed in such a way that the noise level produced in dwelling rooms does not exceed 30 dB (A) or 35 dB (A) in kitchens, and also that more stringent sound insulation requirements than usual may be imposed where commercial enterprises are installed in residential buildings. In the abovementioned 1978 report of the road noise committee, proposals were made concerning further provisions in the building regulations about the siting of rooms sensitive to noise in relation to noisy roads and about the sound insulation of facades.[14]

In concrete terms, many requirements may be imposed in relation to the establishment and use of noisy installations and equipment in pursuance of the Environmental Protection Act. Such requirements may take the form of conditions for the approval of heavily polluting enterprises, etc. under Chapter 5 of the Act, cf. 2.1.2 above concerning the approval scheme. The approval authorities will probably adopt the attitude that the enterprise must comply with specific noise limits, cf. 8.1.3 below, and that it is for the enterprise itself to decide how this can be achieved. In a not inconsiderable number of cases, however, conditions are imposed regarding specific measures for noise reduction, especially covering various forms of noise screening and damping. In addition, the environmental authorities can serve orders concerning remedial measures under section 44, subsection 1, of the Act on enterprises, etc. that are covered by Chapter 5. If an order is violated or if the nuisance cannot be abated by means of an order, the enterprise may be banned in pursuance of section 44, subsection 2. If the enterprise in question has already been approved under Chapter 5, however, orders and prohibitions may only be issued if the pollution substantially exceeds what was adopted as a basis for the approval, cf. section 44, subsection 4. In regard to enterprises not covered by Chapter 5, orders—and, if necessary, prohibitions—may be issued under Chapter 11 of the Environmental Protection Regulations if substantial noise nuisance is involved.

8.1.3 Emission limits

Section 6 of the Environmental Protection Act allows the Environment Minister to issue binding general regulations concerning emission limits for noise. The provisions have not hitherto been used for this purpose but the Environment Agency has issued several guidelines containing

STATIONARY SOURCES

emission standards which have acquired considerable practical importance. Similar standards—binding or recommended—could in theory be laid down for vibration but this would be of little practical significance.

The most important of the Environment Agency's guidelines in the noise sector is *Guidelines No. 3/1974 relating to 'external noise from enterprises'*. As indicated by the title, the guidelines concern external noise from 'enterprises' including public works, etc. and it is expressly emphasised therein that they do not apply to roads, railways, airports, airfields and building and construction operations. Moreover, in practice, they have not been applied to shooting ranges and only to a limited extent to speedway tracks.[15]

The guidelines do not indicate explicit emission limits for noise from enterprises but set out some quality requirements in regard to sound level for different types of areas (corresponding to the normal classification in town plans, etc.) which should not normally be exceeded by external noise from enterprises. The recommended maximum limits are set out as the equivalent continuous corrected noise levels in dB (A) outdoors, the content being as follows:

Area	by day 07.00–18.00	in the evening 18.00–22.00, Sundays and holidays 07.00–18.00 and Saturdays 14.00–18.00	at night 22.00–07.00
1 Industrial areas	70	70	70
2 Industrial areas with ban on operations causing nuisance	60	60	60
3 Areas of mixed residential and industrial development	55	45	40
4 Multi-storey housing areas	50	45	40
5 Areas of open and low-level housing	45	40	35
6 Summer cottage areas	40	35	35

The peak noise levels at night must not exceed the maximum noise levels indicated in the table by more than 15 dB (A).

In cases where the noise from enterprises is largely transmitted through buildings, for example from an enterprise to a dwelling within the same building, the following recommended limits for industrial noise apply measured indoors in a dwelling room with the windows closed:

by day and in the evening	30 dB (A)
at night	25 dB (A)

The guidelines also contain more detailed provisions as to how noise measurements are to be taken, including requirements concerning measuring instruments (reference to specifications in IEC publication 179 (Precision sound level meters)) and conditions of measurement. Furthermore, guidance is given on the assessment of noise measurements, e.g. about taking the variable or intermittent nature of the noise into account. The maximum levels for noise from undertakings are based partly on ISO Recommendation R 1996 (of 1971) and partly on the Pollution Board's publication No. 2, *Industristøj* ('Industrial Noise'), 1971.[16]

The purpose of the guidelines is above all to provide guidance on whether *new enterprises* and extensions to existing enterprises should be approved under Chapter 5 of the Act. The provisions need not be rigorously applied, however, and it is stressed in the guidelines that the table reproduced above should be used in conjunction with the accompanying text. In the first place it does not follow that an individual enterprise should be allowed to emit noise which brings the noise level up to the abovementioned maximum limits. If several noisy enterprises might be established in the area in question, the requirement imposed on the individual installation must be such that the combined noise therefrom does not exceed the noise limits in relation, for instance, to an adjacent area of open and low-level residential development. In the second place, special circumstances may, exceptionally, warrant the emission of noise in excess of the maximum permitted limits. Examples of this are cases where the background noise from traffic, *inter alia*, is substantially higher than the noise from the enterprise or where the enterprise involves a specific beneficial effect. The overall impression from practice is, however, that the standards of the guidelines are observed in the majority of cases and that compliance with the abovementioned noise limits is normally laid down as a condition of approval.[17]

In cases concerning *existing enterprises* too it is assumed by the guidelines that the noise limits should be applied. But here divergences will often be reasonable and the guidelines specially point out that the standards will often have to be applied with a margin of 5–10 dB (A) if the offending enterprise is situated in an industrial area adjoining a residential area. These views in the guidelines are borne out in practice.

As mentioned above, Guidelines No. 3/1974 do not apply to shooting ranges. Consequently the Environment Agency issued recommended standards for *shooting ranges* as *Guidelines No. 2/1979*. It states there that for new shooting ranges the result obtained from a noise measurement according to a more fully described method should not exceed a level that is fixed in each individual case between 65 and 75 dB (A) 'impulse'.[18] In selecting a level there should be taken into account, in particular, the extent of the number of dwellings affected by noise, the shooting activity and the time of day and of year when the range is used. For existing shooting ranges a supplement of 5–10 dB (A) is accepted, likewise according to Guidelines No. 3/1974.

The fundamental rule for *speedway tracks* is that the noise limits in the guidelines concerning external noise from enterprises must be complied with. It has, however, been found difficult to apply these guidelines rigorously to speedway tracks and in *Nyt fra miljøstyrelsen* No. 7/1976 the Environment Agency, without laying down special noise limits, set out a number of more detailed instructions as to the circumstances in which a speedway track should be approved. Compliance with the ordinary noise limits is not normally, in practice, made a condition of approval of speedway tracks but it is sought to ensure an acceptable level of noise by means of requirements relating to situation, screening and use, etc.

8.1.4 Enforcement

In this connection the same applies in all essentials as stated above under 2.1.6 concerning enforcement of the provisions of the Environmental Protection Act on air pollution.

8.1.5 Noise quality targets

8.1.5.1 LEGAL REQUIREMENTS AND RECOMMENDED GUIDELINES

Under current law compulsory quality requirements cannot be laid down for either noise or vibration. On the other hand, the Environment Minister is entitled, under section 8 of the Environmental Protection Act, to issue recommendations concerning noise and vibration, among other things. Such recommended standards have not as yet been established by the Minister but the Environment Agency guidelines mentioned under 8.1.1 and 3 above actually set out quality targets for external noise from enterprises, roads and airfields. As mentioned under 8.1.1, there is every indication that these standards will be incorporated in the regional plans, thus acquiring a binding effect on the authorities' administration of planning and environmental protection legislation.

8.1.5.2 DETAILS OF THE OBSERVANCE OF QUALITY STANDARDS

The quality targets reflected in the Environment Agency guidelines have not been realised in practice. The biggest noise problem is undoubtedly road noise. Research effected by the Planning Executive and the Environment Agency shows that nearly half the population of Denmark is exposed at home to outdoor noise levels from traffic of 55 dB (A) or over, which is unsatisfactory for residential areas. A bare 20% are exposed to an outdoor noise level from road traffic of over 65 dB (A).[19] Approximately 50,000 dwellings—corresponding to 120,000–150,000 people—are affected by noise from aeroplanes and the Environment Agency estimates that 50,000–100,000 people are bothered by train noise.[20] Complaints about noise from industrial and leisure activities, including those about existing enterprises specially, are very common, yet no more detailed information is available about the extent of the noise nuisance. An investigation by the Social Research Institute endeavoured to chart more fully the extent to which the population themselves felt inconvenienced by noise.[21] It emerged from this, among other things, that around 40% of the population felt aggrieved by noise in one form or the other.

It is probably true to state that during the first few years after the environmental reform, from the early 1970s onwards, the main endeavour was to combat freshwater pollution and then air pollution, whereas efforts in the noise sector were more modest. This position seems to be under change in environmental policy; in particular, efforts to reduce the nuisance from traffic noise are being intensified. This is apparent, *inter alia*, from the two reports mentioned under 8.1.1 above concerning road noise and noise from roads and railways dating from 1978 and 1980, respectively. So far, however, the results of this change in course have been very limited.

8.1.5.3 RESTRICTION OF ACTIVITIES IN NOISE-RIDDEN AREAS

If the emission of noise from a given place cannot be avoided—and many extremely noisy activities are unavoidable in modern society—the possibility of restricting the use of such areas for habitation must be considered. The legislation affords various opportunities for applying restrictions of this kind.

It may be done in the first place under *the general planning legislation*; for further details on this see 2.1.1 above. By means of local plans the various types of building development may be separated and between areas for noisy industrial developments and areas sensitive to noise, e.g. residential and summer cottage areas, it is possible to designate areas for less offensive commercial operations. In rural zones the authorities can, through administration of the 'rural zone restriction', ensure that the building of dwellings is permitted only at a suitable distance from noise-producing activities. When the provisions of the planning legislation are applied, the recommended noise levels laid down by the Environment Agency and referred to above under 8.1.1 and 8.1.3 will normally form the basis for deliberations in respect of noise.

There are also means of restricting the availability of areas near roads in the form of the so-called *building lines*, i.e. lines within which no buildings or extensions may be erected or substantial changes made to existing buildings without special permission. Such building lines are sanctioned partly by the Roads Act and partly by the Nature Conservation Act.[22] Under the *Roads Act* decisions concerning the imposition of building lines are reached by the highways executive, which for trunk roads is the Roads Directorate, for highways the county council and for local roads the municipal council. Building lines may be established in relation to both projected and existing roads and may display a mutual distance of up to 100 m for trunk roads and up to 50 m for local roads. Under the *Nature Conservation Act*, a building line of 150 m is generally applicable to trunk roads and highways.[23] Both these types of building lines are imposed for reasons other than the desire to prevent noise nuisance: under the Roads Act in order to allow for road-widening and for traffic considerations and under the Nature Conservation Act to preserve the view from the roads. Nonetheless the building lines have a not insubstantial effect from the noise aspect and may be administered taking this into account.

The 1978 report of the road noise committee acknowledged that it would in principle be possible to achieve satisfactory results as regards keeping clear areas near noisy roads under the existing legislation.[24] The committee nevertheless thought it expedient that this should be supplemented by provisions covering 'noise building lines' proper and suggested that appropriate regulations be incorporated in the Roads Act. According to the committee, these building lines would be based on the regulations which it was simultaneously suggested be issued pursuant to an amended section 10 in the Environmental Protection Act, cf. 8.1.1

above. At this point in time it is not clear whether this suggestion or the committee's other proposals will be carried out, let alone when.[25]

8.1.5.4 INSULATION REQUIREMENT

The 'ultimate' solution to noise problems is to endeavour to reduce the effect of noise by sound insulation and this is the obvious solution in a significant number of cases. In the building regulations issued by the Housing Minister under the Building Act there are, as mentioned above under 8.1.2, provisions concerning the sound insulation of new buildings. These regulations were not, however, like those relating to heat insulation, tightened up during the most recent major revision of the building regulations in January 1977. Legal authority exists for tightening them up and the road noise committee, in its report, suggested certain changes in the regulations with a view to improving sound insulation.[26] In the 1980 report of the traffic noise group a scheme was suggested for the state-subsidised insulation of windows in the existing housing stock.[27] In this context it may also be mentioned that, under the Housing Subsidy Act, state grants for housebuilding are often conditional upon compliance with certain standards of sound insulation.[28]

8.1.6 Individuals' rights

In this connection the same applies in all essentials as set out under 2.1.8 above concerning individuals' rights in proceedings to combat air pollution under the Environmental Protection Act.

Legal proceedings concerning noise and vibration in conjunction with the general law of adjoining properties occur but are very rare. The most important judgment given in recent times concerning noise was the provincial court decision in U 1962 232 where a freeholder's demand for the reduction of the considerable noise nuisance from Kastrup airport and his alternative claim for compensation were rejected. The outcome was based on the fact that the nuisance arose from lawful flying and that the owner must, when he acquired the property, have anticipated a substantial increase in the existing noise. In regard to vibration, there is the Supreme Court judgment in U 1968 84 H where vibrations from driving in concrete piles caused cracks to form in a nearby old building and where the building contractor was ordered, on an objective basis, to pay Dan.Kr. 40,000 in compensation for repairing the damage.[29]

8.2 MOTOR VEHICLES

8.2.1 Construction and equipment

Section 67 of the Road Traffic Act stipulates quite generally that any vehicle shall be so constructed and maintained in such a condition that it may be used without unnecessary risk and inconvenience to others.[30] More specific requirements as to construction and equipment with a view, *inter alia*, to limiting the noise from vehicles may be issued by the Justice Minister pursuant to section 68 of that Act. These requirements are laid down, on the one hand, in a notice relating to the construction and equipment of vehicles[31] and, on the other, in pursuance of section 44 thereof, in a notice of 'detailed regulations for vehicles'.[32] Provisions relating to requirements specially concerned with noise are only to be found in the detailed regulations under subsection 7.05 on noise limits, which is dealt with under 8.2.2 below.

8.2.2 Emission standards

The detailed regulations for vehicles lay down different noise limits for various types of vehicles. For *new standard-type-approved motor vehicles and motorcycles*, the noise measured by *'driven measurement'* must not exceed the following levels:

Passenger vehicles for up to 9 persons incl. the driver	84 dB (A)
Passenger vehicles for more than 9 persons incl. the driver	
permitted gross weight not over 3500 kg	85 dB (A)
permitted gross weight of 3500–5000 kg	89 dB (A)
engine rating of more than 200 hp (DIN)	92 dB (A)
Passenger vehicles for more than 9 persons incl. the driver, more than 5000 kg. gross weight	
engine rating not over 200 hp (DIN)	89 dB (A)
engine rating over 200 hp (DIN)	92 dB (A)
Delivery vans	85 dB (A)
Lorries	
gross weight 3500–12,000 kg	89 dB (A)
gross weight over 12,000 kg	

NOISE AND VIBRATION

engine rating not over 200 hp (DIN)	89 dB (A)
engine rating over 200 hp (DIN)	92 dB (A)
Motorcycles	
cylinder capacity not over 125 cc	82 dB (A)
cylinder capacity of 125–500 cc	84 dB (A)
cylinder capacity over 500 cc	86 dB (A)

New non-standard-type-approved motor vehicles and motorcycles are subject to a series of noise limits from 82 dB (A) to 88 dB (A) measured by *'stationary measurement'*. For *new mopeds* there is a noise limit of 73 dB (A) measured by *'driven measurement'*. *New tractors* must comply with noise limits of 85 and 89 dB (A) measured by the same method according to whether the gross weight is up to 1500 kg or above it.

These noise limits all apply to brand-new vehicles. For *vehicles in use*, the permitted noise limits are 3 dB (A) higher in the case of standard-type-approved vehicles than the result of a 'stationary measurement' recorded in conjunction with the type approval (cf. 8.2.5 below) and for other vehicles 3 dB (A) higher than the noise limits indicated in the detailed regulations.

In the EEC council directives have been issued concerning noise levels for motor vehicles (No. 70/157 of 6 February 1970, as amended by No. 77/212 of 8 March 1977) and for motorcycles (No. 78/1015 of 23 November 1978). These directives are upheld by the Danish rules insofar as the latter do not impose more stringent requirements as to noise than the EEC stipulations. On the other hand, it may reasonably be asserted that the Danish rules have not taken advantage of the opportunities afforded by the directives for tightening up the noise requirements. Both the Environment Agency and the traffic noise group have suggested that these opportunities be grasped.[33]

8.2.3 Maintenance

The requirements concerning the construction and equipment of motor vehicles apply continuously and the owner or person permanently in charge of the vehicle is responsible for its being maintained in such a way that it always complies with the regulations, cf. section 67 of the Road Traffic Act in this connection. At the same time readers are reminded of the difference in noise limits mentioned above under 8.2.2 between brand-new vehicles and those in use. There are no special regulations concerning maintenance with noise in mind.

8.2.4 Use

In this connection readers are referred to the account under 2.2.3 above. To this need only be added that under the Road Traffic Act general speed limits are applicable to motor vehicles. In built-up areas the speed must not exceed 60 km/hour, on motorways 100 km/hour and on other roads 80 km/hour. Mopeds must not be ridden at over 30 km/hour. These speed limits are laid down for reasons of safety and energy conservation but they also have some effect on the noise from vehicles.

8.2.5 Enforcement

On this subject it is on the whole possible to refer to what was said under 2.2.5 concerning enforcement of the provisions of the Road Traffic Act relating to air pollution. Just a few comments should be added on the enforcement of the noise limits indicated under 8.2.2. *Standard-type approval*, which is undertaken by the Ministry of Justice, provides a control that the vehicle complies with the limits laid down by means of a 'driven measurement'. If the noise limit is observed, an additional 'stationary measurement' is effected, the result of which is recorded on the standard-type approval in order to facilitate subsequent checks on vehicles in use by the vehicle inspectorate and the police. Standard-type-approved vehicles may be registered without inspection by the state vehicle inspectorate.

Non-standard-type-approved vehicles are checked at the time of first registration by means of a 'stationary measurement' to see that the noise level does not exceed the limits laid down. The check is undertaken by the state vehicle inspectorate and the measurement result is not recorded.

In practice, standard-type approval is carried out for all taxis and motorcycles, nearly all passenger vehicles and about 50% of vans. Standard-type approval is not normally given to lorries and buses but they are presented for inspection by the vehicle inspectorate at the time of first registration.

The traffic noise group suggested that the control of noise limits be intensified by introducing a requirement that the result of the 'stationary measurement' be recorded for non-standard-type-approved vehicles too, by providing the local police with equipment for stationary measurement and by a greater effort generally on the part of the police and the vehicle inspectorates.[34]

8.2.6 Individuals' rights

In this connection readers are referred entirely to the account under 2.2.6.

8.3 AIRCRAFT

8.3.1 Construction and equipment

Aircraft used for aviation within Danish territory must, under the Aviation Act, hold a certificate of nationality and registration.[35] For Danish aircraft this is issued by the Directorate of Civil Aviation and for foreign aircraft by the authorities in the state concerned. An essential condition for the registration of an aircraft is that it has a certificate of airworthiness, cf. section 9 of the Act. In addition, under this statutory regulation as amended in 1972, the aircraft must comply with any requirements laid down by the Transport Minister with a view to averting noise and other nuisances for persons outside the aircraft. Chapter 3 of the Act contains general provisions on airworthiness certificates, including those relating to the inspection of aircraft and the validity of airworthiness certificates. These regulations were expanded, on the one hand, by a Notice of 1964[36] which contains no rules of special importance in regard to noise, and, on the other, by comprehensive detailed regulations issued by the Directorate of Civil Aviation and not included in Lovtidende (law reports).[37]

8.3.2 Emission standards

Standards relating to limits for noise from aircraft appear in the Directorate of Civil Aviation's detailed directive BL 1/14 which is closely based on ICAO's rules in Annex 16, to which readers are referred.

8.3.3 Maintenance

Aircraft must be permanently maintained in such a way as to fulfil the requirements imposed at the time of registration, cf. section 22 subsection 2 of the Act. If they do not, the certificate of airworthiness is withdrawn. The 1964 Notice referred to under 8.3.1 stipulates *inter alia* that a maintenance manual must be available for every aircraft and that maintenance work may only be carried out by a company or person approved by the Directorate of Civil Aviation.

8.3.4 Use

Pursuant to section 83 of the Aviation Act, the Transport Minister (at the Directorate of Civil Aviation) laid down a number of provisions concerning permitted flight paths. These were designed as far as possible to avoid flying over built-up areas and play a not insignificant part in reducing noise.

By means of an act dating from 1972 civil aviation at supersonic speeds is prohibited over Danish territory.[38] The Transport Minister may, however, make exceptions to this ban in special circumstances. The act represents the realisation in national law of the Nordic Council's recommendation No. 16/1971.

Finally, it should be mentioned that the environmental authorities, when reaching decisions on the approval of airports and airfields under the Environmental Protection Act, may impose conditions concerning permitted times for flying and the number of flying operations. A 1976 case raised the issue of whether the approval authority could also impose conditions to ban the overflying of certain areas or whether requirements of this kind could be imposed only by the Directorate of Civil Aviation under the Aviation Act.[39] The Environment Agency replied to this question in the negative, but the ombudsman found it 'somewhat doubtful' whether an interpretation of the two acts could result in the environmental authorities being prevented from imposing conditions of the type involved and urged the two Ministries concerned to endeavour to clarify the issue further.

8.3.5 Enforcement

The Directorate of Civil Aviation continuously supervises compliance with the requirements for the airworthiness of aircraft and, in accordance with the 1964 Notice, there is a general obligation to report any circumstance of significance for the airworthiness of an aircraft.[40] On the other hand, a further check on noise emission is executed only if relevant changes occur in the engines and construction of the aircraft.

NOISE AND VIBRATION

The rules concerning aircraft are subject to penalties and penal liability is determined according to the general rules of penal proceedings. No legal proceedings have, to our knowledge, been brought on account of noise.

8.3.6 Individuals' rights

In this context the same rules apply by and large as mentioned above under 2.2.6, to which readers are referred.

8.4 SHIPS, INCLUDING HOVERCRAFT

The impact derusting and scraping of ships and engine trials or the like whereby excessive noise is produced must not, under harbour regulations, take place without the consent of the harbour authority. For further details on this matter see 2.3.2 and 3 above.

Under the local by-laws the police may lay down rules for the use of motorboats within a more precisely defined and fairly short distance from areas of coast insofar as is deemed necessary in order *inter alia* to prevent noise which would be a substantial nuisance to other people. This regulation has been used especially to impose restrictions on the use of speedboats within 150 m of the coast.

The regulations mentioned here are subject to penalties and there are no special provisions as to prosecution. In regard to the individual's rights, readers are referred to 2.2.6.

Notes

1. Cf. Environment Ministry circular of 22 February 1977.
2. See e.g. draft regional plan for Århus county of February 1980, pp. 259–69.
3. Cf. Environment Agency publication *Miljøreformen*, 1979, pp. 169–73.
4. Søren Geckler *et al.*, *Fordelingen af levevilkårene* ('Distribution of living conditions'), 1978.
5. Statutory Notice No. 585 of 20 November 1975.
6. Report No. 844/1978, pp. 125–33 in particular.
7. *Op.cit.*, p. 132.
8. *Op.cit.*, p. 126 ff.
9. Cf. Report on combating noise from roads and railways 1980, issued by a working party set up by the Environment Minister in consultation with the Ministers of Housing, Justice and Transport, p. 21.
10. Cf. p. 34 in the report referred to in note 9.

NOTES

11. Statutory Notice No. 381 of 10 June 1969, sections 55–59.
12. Information concerning the aircraft noise load is to be found in the Environment Agency publication *Miljøreformen*, 1979, p. 1970 ff, from which it emerges that approx. 50,000 dwellings in Denmark suffer from aircraft noise, including a good 45,000 from noise from airports.
13. Notice No. 126 of 8 April 1980.
14. Report No. 844/1978, p. 135.
15. Cf. respectively Environment Agency decision of 15 October 1976, *KFE* 1977, p. 80, and the Environmental Appeal Board's unprinted decisions of 9 February and 29 August 1977.
16. Cf. in this connection *Nyt fra miljøstyrelsen* No. 3/1978 which reproduces a report on 'Measurement of external noise from enterprises. Prototype measurements' from the Laboratory of Sound Technology.
17. Further details of administrative practice are to be found in Jens Christensen's *Særlig forurenende virksomhed. Godkendelsespraksis* ('Heavily polluting industry. Approval practice'), 1980, pp. 289–306 (stencilled).
18. *Vejledningen* (guidelines) pp. 39–41.
19. Environment Agency publication *Miljøreformen*, 1979, p. 169.
20. *Op.cit.*, p. 170 ff.
21. Søren Geckler et al., *Fordelingen af levevilkårene*, 1978.
22. For further details see sections 34 and 35 of the Roads Act and section 47 of the Nature Conservation Act.
23. By Act No. 297 of 26 June 1975 to amend the Nature Conservation Act, the provisions of section 47 of the latter concerning road building lines were revoked. The date of taking effect is fixed by the Environment Minister, however, and the amendment to the Act has not yet been implemented. The 1978 report of the road noise committee suggested that this should not be done until 'noise protection lines' had been established with a scope described in more detail by the committee, see report No. 844/1978, p. 113.
24. Report No. 844/1978, p. 115.
25. Cf. *Report on combating noise from roads and railways*, 1980, p. 21.
26. Report No. 844/1978, pp. 111 ff and 135.
27. Report on combating noise from roads and railways, 1980, pp. 27 ff and 55 ff.
28. Statutory Notice No. 388 of 16 July 1979 with subsequent amendements; the Clearance Act too—statutory Notice No. 356 of 3 July 1975—affords an opportunity of ensuring the execution of noise prevention, financed partly by owners and tenants and partly paid for out of public funds.
29. The judgment was commented on by Th. Gjerulff in *U* 1968 B 333 and subsequently aroused considerable discussion in the literature on the law of damages because the compensation appears to have been awarded on an objective basis.
30. Act No. 287 of 10 June 1976, with subsequent amendments.
31. Notice No. 154 of 20 April 1977, with subsequent amendments.
32. Most recently of 24 January 1980 with effect as from 1 April 1980.
33. Cf. *Miljøreformen*, 1979, p. 210, and *Report on combating noise from roads and railways*, 1980, p. 16 ff.
34. Cf. p. 18 in the report referred to in note 33.
35. Statutory Notice No. 381 of 10 June 1969 as subsequently amended.
36. Notice No. 332 of 27 November 1964.
37. Cf. decree No. 42 of 15 February 1964.
38. Act No. 235 of 7 June 1972.
39. Cf. *FOB* 1977 686 in which the environmental authorities' attitude is set out.
40. Cf. section 7, subsection 2, of the Notice.

9
Radioactivity

9.1 NUCLEAR PLANTS

The term 'nuclear plant' ('atomic plant') is used here, in principle, in accordance with the definition in the Paris Convention of 29 July 1960 concerning third party liability in the nuclear energy sector, cf. Art. 1, letter a, ii of this Convention. The definition was in all essentials adopted as a basis in the Atomic Plant Act of 1962[1] and in the Atomic Compensation Act of 1974[2] which contains the current regulations on the subject. In a 1976 Act on Safety and Environmental Protection measures in atomic plants, etc.,[3] which includes the provisions potentially applicable to the establishment and operation of atomic plants, etc., the scope of the regulations was extended somewhat in relation to previous legislation. Thus this Act also covers laboratories, etc. where there are only small amounts of fissionable material and also any activities dealing with the extraction of nuclear fuel and the treatment of non-irradiated natural uranium.[4]

To enable the current rules of law regarding atomic plants to be understood in context, it should be mentioned that there is at present in Denmark only one installation of the type dealt with in the Paris Convention although, at the same time, the establishment of one or more nuclear power stations has been under consideration for a number of years. The sole atomic plant in existence is the state-owned Risø experimental plant which was built under a special Act of 1955.[5] The object of this installation is to conduct research, development work and consultancy operations of importance for the use and control of nuclear energy for peaceful purposes and it is administered by an executive and a management committee.[6]

The question of introducing nuclear power as an element in the Danish energy supply has on several occasions come close to a political decision.

In 1976 the Government planned to settle the matter by presenting a bill but in August of that year the Government informed the Folketing that it did not wish a decision to be reached until further notice. Nowadays Government policy on the matter is that it will present a bill for the use of nuclear power as a source of energy if it is thought that the problems of safety in the vicinity of the plants and of disposal of radioactive waste can be acceptably solved. In the opinion of the Government, not enough light has been shed on these questions to date. It should also be mentioned in this context that a bill concerning the introduction of nuclear power could, at the request of one-third of the members of the Folketing, be submitted to a decisive public referendum in accordance with section 42 of the Constitution; that, if the case arises, such a referendum appears likely and that its outcome is completely unknown.

In the mid-1970s further legislation was passed concerning atomic installations by Act No. 332 of 19 June 1974 on compensation for nuclear damage and Act No. 244 of 12 May 1976 on safety and environmental protection measures in atomic plants. Meanwhile, only the 1974 Act has taken effect, as the date of coming into force of the 1976 Act is to be determined by a special act. It is thought that in this implementing act some standpoint on the question of introducing nuclear power in Denmark may be taken and, as is apparent from the above, nothing definite can be forecast about when the bill for such an act will be presented, let alone whether it will be passed.

This means that, apart from the rules concerning liability in damages for nuclear damage, the current regulations concerning nuclear plants are still to be found in the Atomic Plant Act of 1962, Chapter 2, on which subject readers are referred to subsection 8.2 of the first edition of this publication. Simultaneously, the regulations on safety and environmental protection measures in nuclear plants, etc. are contained in the 1976 Act and will apply as and when atomic installations pose a significant practical problem through the introduction of nuclear power stations. The exposition under 9.1.1–9.1.7 will—apart from the question of liability in damages—be predominantly based on the 1976 Act. In this connection it is observed that this contains only the legal framework and that it is designed to be filled out in many respects by regulations, etc. laid down administratively which do not as yet exist.

Before dealing with the regulations on nuclear plants, it should be pointed out finally that radioactive materials occur in conjunction with these and that the provisions concerning radioactive substances described below under 9.2 apply alongside the special legislation on atomic plants.[7] This legislation merely serves to tighten up numerous aspects of the legal position of working with radioactive materials if this takes place in conjunction with nuclear plants.

9.1.1 Control of the siting of atomic plants

A licensing scheme is intended to ensure that atomic installations are appropriately sited. If a *nuclear reactor plant or an installation for the storage and processing of irradiated nuclear fuel and radioactive waste products* is involved, a licence is granted by the Environment Minister under section 3, subsection 1, and may be granted only if the Folketing has given its approval in a parliamentary resolution. To our knowledge, this is the only case where Danish legislation expressly calls for parliamentary approval of an administrative decision, which shows just how seriously the legislative powers view such decisions. The obvious thing might have been to allow the individual decisions to be taken by act of parliament, as is the case with many major construction projects. In this context the wish to avoid a public referendum on each individual site licence probably played its part, as such referendums may be held only on bills and not on parliamentary resolutions.

In regard to *other nuclear installations* that are covered by the Act,[8] the site licence is issued by the Environment Minister after prior consultation with the parliamentary committee concerned, cf. section 3, subsection 2. This arrangement covers all installations for the production of nuclear fuel, from the crushing of ore or reprocessing of spent fuel to the manufacture of the actual fuel product, and also installations for processing radioactive products, including isotope production, and installations for storing new nuclear fuel or radioactive material.[9]

As a guide to the content of site licences, the Act merely states in section 2, subsection 2, that 'a licence must not be granted when it is deemed questionable having regard to safety or other significant general interests'. A reasonable, expert assessment of the risks associated with nuclear installations is guaranteed to some extent by the Act specifying that the Environment Minister may issue a licence only on the basis of recommendations from the Environment Agency and the Health Board and that the Environment Agency's recommendation must be accompanied by reports on the case from the 'Inspectorate for nuclear plants', which is an institution coming under the Environment Agency, cf. sections 3, subsection 4, and 12, subsection 2, of the Act.[10]

The establishment of nuclear power stations will certainly be conditional on licences, etc. in pursuance of *inter alia* the National and Regional Planning Act, the Municipal Plan Act, the Zone Act, the Environmental Protection Act and the Nature Conservation Act. In this context, section 14 of the 1976 Act provides for the Environment Minister to assume the powers vested in other administrative authorities by this and other related Acts where a nuclear reactor plant is concerned if the location has been approved by the Folketing under section 3, subsection 1. Moreover, the Minister may decide in such cases that a departure may be made from the above-mentioned Acts. On the other hand, the Electricity Supply Act is not affected by the provisions of the 1976 Act. This means that in regard to electricity production with nuclear fuel as the energy source, a licence under the 1976 Act may be granted only to an enterprise that has obtained an exclusive licence from the Energy Minister under the Electricity Supply Act with the approval of the parliamentary committee on energy policy.[11]

9.1.2 Control of the construction of atomic installations

Under the 1976 Act, section 3, subsection 2, the construction and operation of an atomic plant within the meaning of the Act requires a licence from the Environment Minister. The Environment Minister may grant such a licence only after prior consultation with the parliamentary committee concerned. If the Minister has granted a site licence with the approval of the Folketing in pursuance of section 3, subsection 1, of the Act, this lapses if the building and operating licence is not obtained. The more precise requirements as to construction (and operation) of individual plants are laid down mainly in the conditions for its approval, cf. section 5 of the Act, which also enables the more detailed building and operating stipulations to be determined by the Environment Agency and the Health Board within the framework of the conditions laid down by the Minister. Moreover, the Environment Minister may, under section 4, subsection 2, lay down general requirements for nuclear safety and may, in so doing, authorise the Environment Agency and the Health Board to issue the more detailed regulations involved. The building licences will, like the site licences, be granted on the basis of recommendations from the Environment Agency and the Health Board and reports from the Inspectorate for Nuclear Plants, cf. sections 3, subsection 4, and 12, subsection 2.

As to the revocation of licences, the imposition of new conditions and the notification of orders, see 9.1.3 below.

9.1.3 Control of operation and maintenance

Before a nuclear plant is taken into use, an *operating licence* must be obtained under section 3, subsection 2, which is issued by the Environment Minister in accordance with the same rules as building licences, cf. 9.1.2 above. The requirements for the operation and maintenance of the plant will be indicated primarily in the concrete conditions of the licence. It will, if necessary, be possible to issue the operating licence in several stages, for instance, permission for loading fuel and criticality, low-output operation and full-output operation.[12] In pursuance of section 4, subsection 2, the Environment Minister will also, as for construction, be able to lay down *general safety requirements* which may, if necessary, be rendered more specific by the Environment Agency and Health Board.

For the *transport* of nuclear fuel and radioactive material, permission is required for the individual transport under section 3, subsection 5.[13] The Environment Minister may authorise the Health Board to grant such permission. According to statements made by the Minister during discussion of the Bill by the parliamentary committee on energy policy, it will be made a condition of the Health Board's competence that the Board should, before deciding whether to grant permission for transportation, consult the Inspectorate for Nuclear Plants and the Environment Agency.[14]

A licence that has been granted for the siting, construction or operation of an atomic plant or for the transport of nuclear material may *be revoked* under section 6 if important prerequisites for the licence prove to be absent; if conditions imposed are disregarded or if safety considerations or other cogent reasons make it necessary to stop or abandon the plant. Conditions imposed for a licence may be amended and *new conditions* arranged if this is deemed necessary on account of safety or other important general interests, cf. section 5, subsection 3. This means that there are discretionary powers to make conditions relating to atomic plants, etc. more stringent and that the authority to alter decisions relating to licences is considerably wider than in regard to approval under chapter 5 of the Environmental Protection Act.[15]

New requirements for safety and environmental protection measures in conjunction with nuclear installations may also be imposed in the form of *orders*. Under section 7, subsection 1, of the Act, the Environment Agency and the Health Board may give notice of orders necessary to ensure the observance of conditions imposed or which are otherwise deemed requisite on safety grounds, just as they may in an emergency demand that the installation be halted temporarily on safety grounds. The same powers may be exercised by the Inspectorate for Nuclear Plants under section 13, subsection 1.

A final point in this context is that, at the instance of the parliamentary committee on energy policy, provisions concerning a *safety council* were included in the Act. Under section 8 of the Act, a separate safety council has to be set up for every nuclear reactor plant. This council consists of representatives of the staff, the works management, factory inspectorate, the Environment Agency, the local municipal authorities and elected representatives of the local population. The safety council will, upon request or on its own initiative, advise the works management and the authorities with duties concerning nuclear safety in conjunction with the plant. On the other hand, the council has no powers to reach binding decisions on the construction and operation of the plant; it is noted here that a proposal enabling the council to order operation of the plant to stop was rejected in the Folketing. The more detailed regulations concerning establishment and functions of the safety councils are laid down by the Environment Minister. In the comments on the regulation it is assumed that the works' staff and officers will be in the majority on the safety council.[16]

9.1.4 Control of nuclear waste

Licences for the construction and operation of nuclear plants will include conditions relating to the storage and disposal of the nuclear waste. It is noted, moreover, that installations for storing and processing radioactive waste are themselves covered by the licensing arrangements under the Act, cf. section 3 compared with section 1.

9.1.5 Supervision by the polluter

The 1976 Act contains no provisions as to what supervision is to be exercised by the enterprise itself. According to the information supplied by the Environment Minister to the parliamentary committee on energy policy during the passage of the Bill, the owner of the plant will—presumably as a condition of the construction and operating licence—be made responsible for setting up a special organisation for quality supervision. This organisation will be responsible for quality supervision and control, reporting direct to the top technical management of the plant.[17]

9.1.6 Enforcement

Official supervision of nuclear plants is primarily in the hands of the Inspectorate for Nuclear Plants which is, as mentioned, an institution subordinate to the Environment Agency, cf. section 13 of the Act. The Inspectorate may *ex officio* request any information deemed to be important for safety and also demand access to the plant, etc. at any time without prior legal authorisation, cf. section 13, subsection 1, compared with section 7, subsection 1. The same supervisory powers accrue to the Environment Agency and the Health Board. More precise regulations concerning supervision may be laid down by the Environment Minister under section 13, subsection 2.[18]

Infringements of the requirements of the Act and also of conditions attaching to the individual licence are *punishable* under section 16 by *penalties* in the form of a fine, simple detention or imprisonment for up to 2 years but, in the case of contraventions that are inadvertent or of general regulations laid down administratively, only by fine or simple detention. Furthermore, it should be noted that the owner of an installation may be fined even if the infringements cannot be attributed to his deliberate action or negligence and that corporate bodies (companies, etc.) can be fined as such. Of greater practical importance than penalties will probably be the fact that errors or oversights on the part of the enterprise may lead to *administrative sanctions* in the form of more stringent conditions and possibly the revocation of operating licences, etc., cf. 9.1.3 above. In regard to sanctions, moreover, readers are referred to the account under 2.1.6.2.

9.1.7 Individuals' rights

9.1.7.1 THE RIGHT TO OBTAIN INFORMATION

In general, the right of the individual to obtain information is governed by the 1970 Public Access to Documents Act in the same way as within the field of the Environmental Protection Act, see 2.1.8.1 above in this connection.

As regards special rules, the regulation on procedure present in section 14, subsection 7, of the 1976 Act, which was inserted at the insistence of the parliamentary committee on energy policy, should be mentioned.[19] The Environment Minister is hereby directed, in collaboration with the country and municipal councils affected, to hold a *public hearing* about an application for a nuclear reactor plant site before the Environment Agency makes its recommendations thereon, i.e. before either the Environment Agency or the Environment Minister and the Folketing reach a decision on the matter. At this hearing important safety and environmental protection issues involved in dealing with the application are outlined. Moreover, it is assumed that during the hearing the local inhabitants will have an opportunity to put questions to representatives of the authorities that are working on the case.[20] The individual person is thus given access to information well beyond that afforded under the Public Access to Documents Act.

9.1.7.2 RIGHT OF APPEAL

The Environment Minister's decisions and licences cannot be referred to any higher administrative authority. Especially in regard to site licences which are approved by the Folketing under section 3, subsection 1, the possibility of testing them in court is of theoretical rather than practical interest.

The decisions of the Environment Agency and the Health Board concerning conditions and orders, etc. may be referred to the Environment Minister, cf. section 7, subsection 2. Decisions taken by the Inspectorate for Nuclear Plants may be appealed against to the Environment Agency and from thence to the Environment Minister, cf. section 13, subsection 3.

The Act does not state, like the Environmental Protection Act, who is entitled to appeal. This means that the question has to be decided on the basis of the general rules of administrative law concerning the right of appeal. Accordingly, the applicant (the owner of the installation) and 'anyone who has an individual and substantial interest in the outcome of the case' will undoubtedly be entitled to appeal. The latter criterion must presumably be interpreted in accordance with and, at all events, no more narrowly than the provisions of section 74, subsection 1, No. 3 of the Environmental Protection Act[21] and, in particular, the circle of 'neighbours' entitled to appeal must be assumed to be relatively wide.[22] Normally, the county and local councils concerned will probably also have the right to appeal and, it is assumed, often local councils in the municipalities adjoining the one where the plant is situated or is planned.

9.1.7.3 RIGHT TO HAVE VIEWS TAKEN INTO CONSIDERATION

If a citizen has the status of an 'interested party' within the meaning of the Public Access to Documents Act, he or she may, under section 12 of the Act, normally ask for the decision to be deferred until he or she has had an opportunity of expressing an opinion on the matter, cf. 1.7.2 above. This may be said to afford a right, in general terms, to have views put forward taken into account.

Furthermore, the regulation concerning a public hearing mentioned under 9.1.7.1 above must be assumed to mean that the views put forward during the hearing on the part of the local population will 'count' when the authorities are giving further consideration to site licences for nuclear reactor plants. This interpretation may be placed on section 14, subsection 7, last point, and is clearly endorsed by the preparatory work on the regulation.[23]

9.1.7.4 RIGHT TO INSTITUTE AND INTERVENE IN ENFORCEMENT PROCEEDINGS

The individual is only officially entitled to institute enforcement proceedings insofar as he or she has a right of appeal, cf. 9.1.7.2 above. But the supervisory authorities will naturally investigate complaints from private individuals and such complaints may therefore, in practice, result in action being taken concerning an infringement of regulations or conditions.

Unlike the Environmental Protection Act, the Act does not contain provisions to the effect that a complaint can bring about postponement. This means that no 'neighbour' nor, more characteristically, any owner of an atomic installation can, by means of an appeal suspend the execution of enforcement proceedings.[24]

9.1.7.5 RIGHT TO COMPENSATION

The right to obtain compensation for nuclear damage is closely governed by the legislation. The regulations applicable are contained in Act No. 332 of 19 June 1974 which was set into force by Notice No. 467 of 16 September 1974. The content of the regulations accords with the Paris Convention of 29 July 1960 with supplementary convention of 31 January 1963 and the Convention of 17 December 1971 concerning civil law liability in conjunction with the transport of nuclear material by sea and is based in all essentials on a 1968 commission report.[25] Since the regulations accord with the contents of international conventions, the provisions will not be further examined here.[26]

It should be added that at present Denmark has no plans to ratify two other conventions concerning compensation for nuclear damage—the international Convention of 25 May 1962 on the liability of owners of nuclear ships and the Vienna Convention of 21 May 1963 on civil law liability for nuclear damage.[27] The Atomic Compensation Act of 1974 is nevertheless drafted in such a way that it would be relatively easy to make the amendements to Danish regulations needed to be able to ratify these two treaties. Thus it is laid down in section 39 of the Act that the Justice Minister can lay down regulations concerning the rules under which compensation is to be paid for nuclear damage caused in an accident involving a nuclear-powered means of transport. This provision can provide the authority for the administration to issue compensation regulations answering to the 1962 Convention and may also be drawn on to issue regulations in accordance with a bilateral agreement in conjunction with a visit to Danish territory by a nuclear-powered ship.[28]

9.2 RADIOACTIVE SUBSTANCES

9.2.1 Control over siting

Any enterprise, etc. wishing to produce or utilise radioactive substances must, of course, keep within the frameworks laid down by general planning legislation, see 2.1.1 above in this connection. Yet these are seldom, if ever, directly relevant to radioactive substances.

On the other hand, there is some special legislation dating from 1953. Act No. 94 of 31 March 1953 on the Use, etc. of Radioactive Substances specifies in section 1 that radioactive substances of any kind may only be produced, imported or held if permission to do so has been granted by the Health Board. The Environment Minister may, nevertheless, make exceptions to this general prohibition. This was done in Notice No. 127 of 4 May 1953 which exempts from permission certain substances, certain goods (e.g. watches) and certain institutions (of a scientific nature, in particular). Under the abovementioned Notice the Health Board may, moreover, grant importers and makers of instruments containing radioactive material permission for their sale and further transfer without the trader having to apply for permission to the Health Board in each individual case.

9.2.2 Control of storage and use

The 1953 Act does not indicate any more specific requirements for the storage, etc. of radioactive substances but empowers the Minister to issue regulations concerning safety measures in conjunction with the importation, production, use, storage, transport and disposal, etc. of radioactive substances. The powers were exercised by virtue of a Notice of 15 June 1955 which has now been superseded by Notice No. 574 of 20 November 1975.

The Notice requires in general that the safety measures, unless otherwise specified in pursuance of the 1953 Act, shall be taken in accordance with the guidelines recommended at any time by the ICRP (International Commission on Radiological Protection). The Health Board acquaints users of radioactive substances with these recommendations by distributing guidelines.

In addition, the 1975 Notice contains a few specific provisions concerning storage and keeping (Chapter II), production, processing and use (Chapter III), importation and transport (Chapter IV) and disposal, etc. (Chapter V). The provisions of the Notice on these points are quite concise and substantially more summary than the provisions contained in the previous Notice of 1955. This reduction in content of the Notice is bound up with the 1975 Notice containing on all the points mentioned a subdelegation of powers enabling the Health Board to issue more detailed regulations not merely in the individual instance but also of a general nature. The subdelegation of powers to issue regulations under a Notice generally gives rise to misgivings about the constitutional aspect when it is not expressly stated in the statute that the Minister may transfer his powers to others. When, as in the present case, regulations of a specifically technical nature are involved, however, subdelegation has in practice also occurred previously without being expressly sanctioned by statute.

The Health Board's general regulations, which are not included in *Lovtidende*, are quite comprehensive and detailed in many respects and it would be beyond the scope of this account to reproduce them more fully.

As regards the storage of radioactive substances, the Notice stipulates *inter alia* that the user of radioactive materials must exercise care that no-one is exposed to health risks through direct radiation or the absorption of radioactive substances into the body via the airways, digestive tract or skin. Accordingly the radiation doses indicated by the ICRP must not be exceeded and it must be a constant endeavour to see that the radiation doses lie substantially below these. From the other statutory regulations concerning storage, it should be mentioned that places used for storing and keeping the substances must be clearly marked with a notice warning of radioactivity and secured against theft, fire and damage by water.

Places where radioactive substances are produced, processed and used must also be clearly marked with notices warning of radioactivity. Beyond this, it is left to the Health Board to lay down regulations concerning the manufacture, processing and use of radioactive substances.

9.2.3 Control of packaging and transport

On this point the 1975 Notice in essence only empowers the Health Board to issue directives. These may cover packaging, marking and shielding. For transport by sea, rail and air, regulations concerning radioactive substances are laid down, however, by the Industry Minister, the Minister of Public Works and the Directorate of Civil Aviation, respectively.

9.2.4 Control of disposal

The 1975 Notice empowered the Health Board, on the one hand, to issue general regulations for dealing with radioactively polluted packaging, waste, waste water and extracted air and, on the other, to lay down special provisions concerning these in individual cases. For industrial undertakings it is usually specified that wastes must be returned to the supplier, a foreign undertaking or the Risø Experimental Station (see 9.1 above) or to the latter's installations for collecting and processing radioactive waste.

9.2.5 Supervision on the polluter's part

There are no general regulations on this point but requirements may be imposed as conditions in specific cases. The polluter will, moreover, generally have an interest in exercising supervision in order to avoid committing a punishable offence.

9.2.6 Quality targets

See 9.2.2 above concerning the ICRP's guidelines.

9.2.7 Enforcement

Supervision of the observance of the Act on radioactive substances and the Notices, etc. issued in pursuance thereof is exercised by the Health Board through the Radiation Hygiene Institute which forms part of the Board's organisation. For undertakings that are covered by the Working Environment Act, however, the supervision is carried out by the local factory inspectorate which receives copies of licences issued by the Health Board. The Health Board carries out neither routine nor random inspections after the licence has been granted but occasionally looks at various types of enterprises involving radioactive substances in order to keep abreast of current practice in conjunction with their use, etc.

The Health Board's decisions concerning licences for and requirements relating to the possession, etc. of radioactive substances may be referred to the Environment Minister but this has only occurred in extremely few cases. An appeal by an undertaking against a Health Board order would not have a suspensive effect.

Infringement of the statutory rules and of regulations laid down administratively may result in fines and the proceedings are conducted in accordance with the general regulations on criminal proceedings. Infringements occur but normally the person concerned puts matters right when cautioned by the supervisory authority and criminal proceedings are virtually never instituted as a result of information passed on by the Health Board. This is presumably due *inter alia* to the use of radioactive substances being subject to a licence and to the fact that this may, in cases of gross or repeated disregard of regulations and orders, be revoked in accordance with the general principles of administrative law.

9.2.8 Individuals' rights

In conjunction with the regulations on radioactive substances dealt with above, third parties, i.e. persons other than the owner, user or transporter concerned, will scarcely ever have the status of 'interested parties' (cf. 1.7.2 above on this), which is very important for third party rights.

Individuals are therefore entitled to obtain only information concerning the contents of documents under chapter 1 of the Public Access to Documents Act; for further details, see 1.7.2 above. Moreover, an outsider has no right of appeal against administrative decisions to superior administrative authorities, to have his views taken into consideration or to suspend enforcement proceedings. Compensation for damages may be claimed under the general rule of *culpa*, cf. 1.7.6 above, but no instances of proceedings for compensation for damage resulting from radioactive substances are known.

Notes

1. Act No. 170 of 16 May 1962.
2. Act No. 332 of 19 June 1974.
3. Act No. 244 of 12 May 1976.
4. Cf. section 1 of the Act together with H. J. Stehr in *Karnovs Lovsamling*, 9th ed., p. 1528 *et seq.*
5. Act No. 312 of 21 December 1955.
6. See now Act No. 194 of 28 April 1976 on measures concerning energy policy, Chap. 3.
7. Cf. section 15, subsection 1, of the 1976 Act.
8. Cf. section 1, Nos. 1 and 2 of the Act and the introduction to 9.1 above.
9. There are nevertheless exempted from control under section 15, subsection 2, of the Act enterprises that import, process and store radioisotopes for use in industry, commerce, agricultural technology, medicine or for scientific purposes and also radioactive products consisting of radioisotopes to be used for the said purposes.
10. Cf. in this connection *Folketingstidende* 1975–76, appendix B, col. 972 *et seq.*
11. Cf. section 15, subsection 1, of the 1976 Act compared with the Electricity Supply Act, Act No. 54 of 25 February 1976, sections 4 and 5.
12. Cf. *Folketingstidende* 1975–76, appendix B, col. 1022.
13. It is noted in this connection that the Act does not cover radioisotopes, etc. for use in industry, commerce, agricultural technology and for scientific purposes, cf. note 9 above for further details.
14. *Folketingstidende, op.cit.*, col. 1026 et seq.
15. Cf. 2.1.2.2 above.
16. Cf. *Folketingstidende, op.cit.*, col. 971.
17. For further details see *Folketingstidende, op.cit.*, col. 1022.
18. Cf. in this connection *Folketingstidende, op.cit.*, cols. 1019–23, where the Environment Minister gave a brief statement on some preliminary deliberations concerning the monitoring of safety conditions in atomic plants.
19. Cf. *Folketingstidende* 1975–76, appendix B, col. 967 *et seq.*, 974, 983–85, 1029 *et seq.* and 1037–39.
20. Cf. *op.cit.*, col. 974.
21. For further details, see 1.7.3 above.
22. Cf. the Environment Minister's statements to the parliamentary committee that dealt with the Bill, *Folketingstidende, op.cit.*, col. 1039.
23. Cf. *Folketingstidende, op.cit.*, col. 1938.
24. In regard to the general rules of administrative law concerning the suspensive effect of appeals, see Ellen Margrethe Basse, in *Forvaltningsret, Almindelige emner*, 1979, p. 305 *et seq.*
25. Report No. 482/1968.
26. There is an earlier article on the subject of liability in damages for nuclear damage by Bernhard Gomard in *U* 1958 B, pp. 197–235.
27. Cf. *Folketingstidende* 1973–74, appendix A, cols. 760–64 with references to the 1968 report.
28. Cf. in this connection, Notice No. 243 of 29 July 1964 concerning liability in damages for the nuclear-powered American vessel n/s 'Savannah' in accordance with an agreement of 2 July 1964 between the USA and Denmark.

10
Product Controls

In the preceding Chapters various forms of environmental nuisance have been classified according to the nature of the pollution or the object of pollution. In several contexts the legal requirements concerning products with a view to preventing or reducing pollution are referred to there. All environmental protection regulations relating to products could be assigned to Chapters 2–9, but for purposes of presentation it is expedient to collect certain product regulations together in a separate section. The main reason is that special legislation with an environmental protection slant exists covering chemical substances and products, including poisons and pesticides. To provide an overall view, references are also given to the most important passages in Chapters 2–9 where product regulations have been dealt with previously. This does not apply, however, to rules concerning means of transport, in which connection see 2.2 and 2.3 and 8.2–8.4 above.

The following account covers only legislation designed, at least in part, to protect the external environment. Also included are regulations with the prime object of protecting human health from immediate danger—chiefly in the form of poisoning—if they also have a broader environmental protection aim. On the other hand, legislation concerned solely with human health, with the protection of workers and employees and buyers of goods is not taken into account. Hence neither the drugs, working environment nor food legislation is incorporated in the exposition.

Until 1 October 1980 the most important environmental laws relating to products were the Poisons Act and the Pesticides Act.[1] From the abovementioned date these two Acts were replaced by Act No. 212 of 23 May 1979 relating to chemical substances and products designated below as the '*Chemicals Act*'. This Act covers the matters that were previously governed by the Poisons Act and the Pesticides Act but its rules also incorporate legal authority for the Minister of the Environment to issue regulations to restrict or prohibit the sale, importation and use of substances that previously existed under section 7 of the Environmental Protection Act and in other respects contains controls or opportunities for control without counterpart in the preceding legislation. The Chemicals Act is one of the most significant reforms in the environmental protection sector since the adoption of the Environmental Protection Act in 1973 and contains machinery for very intensive controls, whose scope can only be seen in bare outline at present. Behind the reform there lies first and foremost a growing awareness especially of long-term harmful or potentially harmful effects from the current widespread use of chemical substances and products.[2] In order to be able to check these, it has been necessary to provide new legislation which, on the one hand, allows for more ready intervention than formerly and, on the other, enables information concerning chemical substances and products to be collected with a view to further controls, if necessary. Among the considerations leading to the reform was the fact that compliance with the EEC Council Directive of 18 September 1979 (831/EEC) concerning the 6th amendment to Council Directive 1967/548 involved a change in Danish legislation.[3] In the Chemicals Act an attempt was made to frame a collective new set of rules in the sector. However, the division of the former legislation between poisons, etc. and pesticides has still left its mark on the current provisions, as is reflected below.

Product regulations based on environmental considerations also exist outside the range of the Chemicals Act. It is noted, in particular, that section 6 of the *Environmental Protection Act* affords considerable scope for the Minister of the Environment to issue product regulations, although these legal powers have hitherto been used only to a very limited extent.[4]

We start below (10.1) with a general review of the Chemicals Act, including its notification scheme. Then follows under 10.2 a discussion of the regulations relating to dangerous chemical substances and products and, under 10.3, of pesticides. These subsections form the main part of Chapter 10 of the survey. The succeeding brief subsections refer to more specific product controls under the chemicals legislation (10.4 and 10.5), on the one hand, and in pursuance of other legislation (10.6–10.9), on the other.

10.1 GENERAL ASPECTS OF THE CHEMICALS ACT

The *sector* covered by the Act is 'chemical substances and products'. 'Chemical substances' are defined as basic substances and compounds thereof as they occur naturally or are produced industrially. By 'chemical products' are understood both solutions and solid, liquid or gaseous mixtures of two or more chemical substances.[5] The Environment Minister may rule, moreover, that products of which the properties are determined by their content of viruses, bacteria, fungi or other microorganisms, together with special organisms that are used instead of chemical substances and products for purposes within the scope of the Act, shall be wholly or partly covered by its provisions, but in the case of the last group only by Chapter 7 thereof.[6] These regulations may for instance, be applied to biological pesticides. The Minister is similarly empowered to exempt specific chemical substances and products and otherwise, in special cases, to permit deviations from the Act.[7] It should also be pointed out that the Act, under section 8, does not apply to chemical substances and products that are produced abroad and are only conveyed across Denmark or stored here as goods in transit. Nor does the Act, apart from the notification rules in Chapter 3, apply to substances and products being exported, unless the Minister of the Environment decides otherwise.

The *means of control* under the Act may be divided into the following categories:

(1) a notification scheme;

(2) general regulations concerning the classification, packaging, labelling, storage and sale of dangerous substances and products;

(3) powers to impose restrictions through regulations issued administratively on the stocks and use of chemical substances and on the composition of chemical products;

(4) an approval scheme;

(5) powers to intervene with concrete orders and prohibitions.

Points (2) and (4) are dealt with below under 10.2 and 10.3, respectively.

Generally speaking, these means of control are subject to two requirements. First, they must be applied in accordance with the aims of the Act, i.e. to prevent damage to health and to the environment which may stem from chemical substances and products, cf. section 1. Second, when deciding what measures to apply, a balance must be struck between, on the one hand, the damage which the substance or product may cause to the environment and, on the other, the technical and economic consequences, including the cost, of such steps both for society and for the individual, cf. section 2, subsection 2.[8] As provided by the Act, it appears that such 'balancing' should be undertaken only if it is a question of 'environmental damage', not if there is a risk of 'damage to health'. The comments in the bill on this regulation state that where a risk to human health is involved, weighing up of the type outlined does not seem practicable.[9] If a very low degree of probability of disease being caused is involved, it is doubtful whether such weighing up can be described as impracticable but, as the Act is worded, it is uncertain whether it is authorised.

Rules concerning *notification* occur in Chapter 3 of the Act. With these are associated a Notice[10] and also Guidelines No. 2/1980 from the Environment Agency. Under sections 11 and 12 of the Act, a 'new chemical substance' may not be sold or imported until the producer or importer has executed investigations to clarify its effects on health and the environment and duly notified the Environment Agency. A substance is regarded as 'new' if it has not been sold in or imported into Denmark prior to 1 October 1980. The duty to notify concerns only chemical substances, but is irrespective of whether it is intended to sell or import the substance as such or as a component of a chemical product. Similar notification regulations apply to chemical substances that were sold in or imported into Denmark before the abovementioned date if they are thereafter sold or imported for a substantially different application or in a substantially higher quantity. Furthermore, section 18 empowers the Environment Minister to also bring existing substances, i.e. substances that were sold and imported prior to 1 October 1980, under the regulations concerning notification, etc. For the time being it was considered unworkable for both enterprises and authorities to have the scheme cover more than new substances but at the same time it was thought desirable to provide for the gradual extension of its scope.

Under section 7 of the Act the scheme applies neither to chemical waste[11] nor to chemical substances that are exclusively sold, imported or used as or in medicines and as additives in foods or feedingstuffs. Moreover, section 6 of the Notice exempts

(1) substances which are sold or imported in quantities of less than 1 tonne per annum per producer or importer;

(2) polymers containing less than 2% in bound form of a new monomer;

(3) substances in the research and development phase which are sold or imported for industrial research and development purposes at the hands of a limited number of enterprises.

If the new substance is produced in another Common Market country, it does not have to be notified in Denmark if it is notified in the producer's native country in accordance with Council Directive 1979/831, cf. section 9 of the Notice.

It may perhaps in some cases prove difficult for the seller or importer of a chemical substance to discover whether it is 'new' in Denmark, as there is no list of substances that were sold or imported here prior to 1 October 1980. It is, however, incumbent on the seller or importer to make the inquiries necessary to ascertain whether a substance is new and the responsibility for wrong assessment is, in principle, his or hers.[12] This is one of the substantially new points in the Act compared with the earlier Poisons Act. The controls now cover substances that are additional to those on an official list. If one wishes to sell or import a substance, there is a far-reaching obligation to investigate it or have it investigated, to ascertain its effects on health and the environment and to assess whether it is new to the Danish market or not.

The notification must contain various details and proposals for the classification and labelling of the substance, if applicable, cf. section 13 of the Act. An annex to the Notice sets out more fully the fairly comprehensive details, etc. which must accompany the notification. The substance or product may not be sold until at least 45 days after the notification has been received by the Environment Agency. If the notification complies with the requirements of the Notice, the applicant will be informed accordingly before expiry of the 45-day period. If the notification was inadequate or incorrect, the Environment Agency can and will order that the substance or product may not be sold until at least 45 days after receipt of the information which the notification should contain.[13]

Notifications must be submitted in duplicate, one copy being assigned to an EDB register which is kept by the Working Environment Institute under the Ministry of Labour and belongs jointly to this Ministry and the Environment Ministry.[14]

The Danish notification rules fulfil the requirements of the EEC Directive of 18 September 1979 but are in many respects more stringent than its provisions. First, the duty of notification takes effect in accordance with the Danish regulations for all substances which were not sold or imported into Denmark prior to 1 October 1980, regardless of whether they were sold in or imported into another EEC country before that date. Second, the Directive fixes 18 September 1981 as the latest date for a notification scheme to come into force whereas the Danish Notice took effect on 1 October 1980. Owing to this difference the Environment Agency has, in the intervening period, been more ready than otherwise to waive the requirements relating to investigation of the substance in question.[15]

As mentioned above, the Chemicals Act also makes provision for far-reaching controls through *general regulations issued administratively*. Section 30 combines the previous authorities in section 15 of the Poisons

Act and section 7 of the Environmental Protection Act and empowers the Environment Minister to draw up regulations for restricting or prohibiting the sale, importation and use of chemical substances and their purity. The Notices issued before the Act came into force in pursuance of section 15 of the Poisons Act and section 7 of the Environmental Protection Act concerning the degradability of washing and cleaning preparations, the importation and use of PCB and PCT, the use of poisons and substances dangerous to health for specially indicated purposes and the restriction of dioxins in pentachlorophenol, etc. are upheld by means of section 67 of the Chemicals Act.[16]

Under section 31 the Minister can issue directives concerning the composition of chemical products, including the extent to which chemical substances may also occur in the product as impurities. This rule corresponds by and large to section 15 of the Poisons Act which provided the legal authority for the Notice of 2 January 1979 still in force relating to cosmetics.[17] Section 31 also empowers the Minister to issue regulations concerning the microbiological purity of the product.

Section 32 enables the Minister to stipulate that, for special purposes or in specific products, only chemical substances that are specially permitted for such purposes or in such products may be used. This introduces for the first time in Danish environmental protection legislation clear and definite authority to implement a 'positive list system' in more precisely defined sectors, i.e. an arrangement whereby only substances that are expressly permitted, by being included in a list, may be used for the purpose in question.

The Chemicals Act also provides for the issue of *orders and prohibitions*. These may be issued firstly under the section 30 referred to above in respect of the sale, importation or use of a chemical substance. The regulation may, for instance, be invoked if the Environment Agency upon receipt of notification of a substance realises that concrete intervention is needed, e.g. in the form of a temporary prohibition, until further investigations have been completed. Then the supervisory authorities—namely, the Environment Agency, the Chemicals Inspectorate, which is an independent body under the Environment Agency, and the local councils[18]—may, under section 48, issue orders and prohibitions in the event of violation of the Act or regulations issued in pursuance thereof. The scope of such powers is not clear from the wording of the provision or the preparatory works. It may probably be assumed that they cannot authorise the imposition of requirements not sanctioned by other regulations in an Act or Notice and that they only sanction orders defining these requirements more precisely and prohibitions where these are not complied with.

An extremely important element in the Chemicals Act is an extensive *duty to inform and investigate* for producers and importers of and dealers in chemical substances and products. They are not merely bound to provide information and effect investigations prior to notification and applications for approval. Upon request, any producer, importer or dealer must supply the information concerning the substance or product, including financial and accounting details, which is requisite for the administration of the Act, cf. section 39. Under section 40, moreover, producers and importers must, when requested, carry out or have carried out on their own account investigations into the effects of the substances or products 'if there is reason to assume that a substance or product has effects that are dangerous to health or harmful to the environment'. This applies even to a substance that has already been duly notified, cf. section 41.[19] In particular, this duty to investigate extends further than the duty to investigate under section 52 of the Environmental Protection Act. The regulations may be seen as a far-reaching and consistent application of the principle of 'the polluter pays'.

Information under the Chemicals Act is subject to the general rules of the Public Access to Documents Act concerning *access to documents*, cf. 1.7.2 above. More precisely in cases concerning chemical substances and products, questions concerning 'trade secrets' will often arise and the exemption rules of section 2, subsection 1, point 2, of that Act will to a not insignificant extent afford a basis for withholding technical and trade information of vital importance to the producer or importer.[20a]

Decisions taken in the first instance by the Chemicals Inspectorate and the local councils may, under general unwritten rules of *appeal*, be referred to the Environment Agency. The latter's decisions may, under section 45, be made final by the Minister on behalf of the administrative authorities and that has, in the main, been done in the complementary Notices. Under section 56 of the Act, decisions concerning orders and prohibitions pursuant to section 30 and decisions on approval, etc. under Chapter 7 may, however, always be appealed against to the Environmental Appeal Board. The competence of this body may, by means of Notice regulations, be extended to other cases under Chapter 3 and 6.[20b]

Infringements of the regulations of the Chemicals Act and directives issued in pursuance thereof are *punishable* by fine, simple detention or imprisonment for up to one year. In regard to criminal responsibility, readers may in essence be referred to the account under 2.1.6.2.[21] Appeals generally have a *suspensive effect* but the appeal or supervisory authority may expressly decide otherwise, cf. section 54.

10.2 DANGEROUS CHEMICAL SUBSTANCES AND PRODUCTS

Chapter 4 of the Chemicals Act contains provisions relating to the classification, packing and storage of dangerous substances and products, etc. which are supplemented by the rules in Chapter 5 concerning the sale of poisonous substances and products. The clauses in Chapter 4 of the Act are developed by Environment Ministry Notice No. 408 of 17 September 1980, as amended by Notice No. 147 of 16 March 1981, together with the Environment Agency's Guidelines No. 2/1981. These regulations accord with EEC Council Directive 1967/548 with subsequent amendments.

The new rules in this sphere correspond largely with the previous regulations contained in and pursuant to the Poisons Act.[22] A change worth special mention is that the obligation on the part of the producer, importer and dealer is no longer unilaterally linked with an officially prepared list of dangerous substances. Such a document is still brought out by the Environment Ministry and is revised about once a year, cf. most recently Environment Ministry Notice No. 147 of 16 March 1981 with annex. If a substance is covered by this list, it must continue to be classified and labelled in accordance with the indications therein. If the substance is not on the list, however, it is the producer's responsibility and task to assess the dangerousness of the substance and, if necessary, to see that the necessary classification and marking take place. It should also be mentioned that the labelling obligations have been extended somewhat.[23]

The fundamental provision is section 19 of the Act, in accordance with which every producer or importer of a chemical substance or product must, before selling or importing it, must obtain such information about the properties and effects of the substance or product as to enable him to classify, package and label it in accordance with the regulations in the Act and Notice.

Classification The Notice lists the classes of hazard as (a) highly poisonous, (b) poisonous, (c) harmful to health, (d) caustic, (e) local irritant, (f) explosive, (g) combustible, (h) highly flammable, (i) readily flammable and (j) flammable. At the same time Annex 1 to the Notice sets out more detailed classification criteria for dangerous chemical substances and products. If the substance is included in the abovementioned list, the classification problem of the producer and importer is thereby solved. Otherwise he or she must, on the basis of the criteria set out, decide on the classification, packaging and labelling associated therewith. In many cases a perusal of existing reference literature or an expert opinion will suffice for this purpose[24] and there is no provision for investigations similar to those required under the regulations concerning notification of new substances. A producer or importer who has classified a chemical substance as dangerous must, when the substance is put on the market, inform the Environment Agency of its name and also the hazard class and what indications of risk and safety instructions have been chosen, cf. section 7 of the Notice.

The Notice sets out in Chapter III a number of requirements in regard to *packaging*. The chief purpose of these is to ensure that the packaging is solid and strong and that no possibility of confusion with harmless substances and products can arise. It is specially noted that packages with a capacity of 3 litres and over which are on sale retail to the public must, to a large extent, be provided with child-resistant closures.[25]

Rules on *labelling* appear in Chapter IV of the Notice. The labelling must include the trade name of the substance or product, the name(s) of the dangerous substance(s), the name and address of the producer, the hazard class of the goods and more specific danger symbols. In addition, the labels on the goods must, as a rule, carry standardised risk indications and safety instructions. This does not apply, however, to substances and products that are not on retail sale to the public or if the capacity of the pack is below 125 ml and the substance or product is classified as a local irritant, readily flammable, flammable or combustible.[26]

If a chemical substance or product is classified as poisonous, i.e. as 'highly poisonous' or 'poisonous', its *sale* is governed by special provisions in Chapter V of the Act and sections 23–37 of the Notice.[27] These provisions chiefly regulate who may sell and buy poisons. Poisonous substances and products may be bought and sold by a number of enterprises without special permission; see section 25 of the Act for further details. The retail sale of poisons may, however, be effected only by pharmacists and by traders who have obtained permission to do so from the Chemicals Inspectorate on conditions laid down more specifically. Poisons may be sold only to private individuals if they are over 18 and against delivery of a special requisition which must be endorsed by the police. The requisitions must each be provided with a serial number and kept in numerical order for 5 years.[28]

Under section 23 of the Act, the Environment Minister may, in general, issue regulations for the *storage* of dangerous substances and products. More detailed provisions of this kind are to be found in Chapter VI of the Classification Notice, which nevertheless relates chiefly to the storage of poisons.[29]

Rules regarding the *transport* of dangerous substances and products are issued by the special authorities who are responsible for the type of conveyance involved.[30]

The Environment Agency is the highest *appeal authority* in relation to decisions under the Classification Notice.

10.3 PESTICIDES

Pesticides are understood to be chemical substances and products which are used to combat plant diseases, weeds and pests. Chapter 7 of the Chemicals Act established an approval scheme for substances and products used for a purpose mentioned in an annex to the Act. At present this schedule covers only more precisely specified pesticides but the Environment Minister can amend the schedule, which includes extending it. In regard to pesticides as defined in this schedule a general Notice, No. 410 of 17 September 1980, was adopted concerning pesticides which established under Chapter 7 of the Act, on the one hand, more detailed regulations for the approval of pesticides and in pursuance of Chapters 4–6, on the other, a number of other rules on classification, labelling, sale, storage and use. In conjunction with these regulations, the Environment Agency also issued Guidelines No. 3/1980 relating to pesticides. On some points these rules are supplemented by special Notices on pesticides; aerial spraying, for example, is also covered.

Before this reform of the legislation on chemical substances and products the regulations concerning pesticides appeared in a special Act of 1961[31] with associated Notices. The new set of rules is essentially a development of the previous state of law but in many respects the current provisions imply an extension and intensification of the controls. *Inter alia*, the Chemicals Act entails a more comprehensive definition of the term 'pesticides' as it includes certain wood preservatives, deterrent agents, certain plant-growth control agents, agents for combating slime-forming organisms in paper pulp and agents for combating algal growth, etc.[32]

Under the Act and Notice pesticides must, prior to sale, importation or use, be approved by the Environment Agency.[33] The Notice stipulates more precisely what information must accompany the application. Approval is given with more detailed conditions relating to the content, quantity, sale, importation, packaging, advertising and marking of the substance or product. The approval once granted may, under section 48 of the Act, be *withdrawn* if the conditions thereof are violated or if fresh information is deemed to make this necessary. Withdrawal will take place if it is found that pesticides involve serious risks to health or the environment.

PRODUCT CONTROLS

Whilst the regulations in and pursuant to Chapter 3 of the Act relating to notification apply to pesticides as well as to other chemical substances and products, the Minister has issued in the Pesticides Notice special rules under section 29 concerning *classification, packaging, labelling, sale and storage*. It is merely noted that under these regulations both highly poisonous and poisonous pesticides may be sold only by the holder of a licence for the preparation concerned and by persons who have, on more fully defined conditions, obtained permission from the Chemicals Inspectorate to do so.[34] Such permission may be withdrawn by the same authority. The relevant conditions are not mentioned in the Notice but may be assumed to coincide with the reasons for withdrawal affecting approvals, which are set out in section 38 of the Act (see above) and which, in essence, express the general unwritten rules of administrative law concerning the annulment of administrative decisions of the type involved.[35] The rules covering the right to sell pesticides differ, moreover, according to whether they are 'highly poisonous' or 'poisonous'.[36] The scope of lawful buyers is defined in more detail in the Notice, in particular relative to the group that may use poisonous pesticides; in this connection, see below.[37]

Pursuant to section 30 of the Act, the Notice also contains provisions relating to the *use* of pesticides, stipulating *inter alia* that 'highly poisonous' pesticides may be used only by a person who has received special permission to do so from the Chemicals Inspectorate together with his/her assistants. Similarly, 'poisonous' pesticides may be used by someone who has a professional interest in the use of the preparations in his/her own business, or anyone who, in the course of business, carries out pest control operations for others and also their assistants.[38]

As regards *aerial spraying* of pesticides, there is a separate Notice, the most recent being Notice No. 185 of 15 April 1981. This stipulates *inter alia* that the distribution of pesticides from aircraft may be undertaken only by enterprises that are licensed to do so by the Environment Agency, on specific conditions if necessary, and that the distribution must only be carried out by specially trained pilots who have obtained an aerial spraying licence from the Environment Agency. The Notice also contains provisions as to which pesticides may be distributed by aircraft, the requirements concerning minimum area and distance and the reporting obligation. By way of special rules governing pesticides, there is also a Notice issued pursuant to the Pesticides Act and still in force concerning the sale and use of *mercurial pesticides* for dressing grain and seed.[39]

10.4 USE OF POISONS AND SUBSTANCES DANGEROUS TO HEALTH FOR SPECIAL PURPOSES

By means of Notice No. 349 of 16 June 1977, the Environment Minister issued, in pursuance of the Poisons Act in particular, regulations concerning the use of poisons and substances dangerous to health for a number of specially stipulated purposes. The Notice was subsequently amended in 1979 and 1980.[40] The regulations are upheld by the Chemicals Act, section 30 of which empowers the Minister to amend and extend the scope of the Notice.

The provisions of the Notice relate to a wide variety of circumstances. The sole common feature is that they set out certain restrictions on the right to use dangerous substances in connection with purposes more precisely enumerated.

It would go beyond the bounds of this account to discuss the individual regulations in the Notice in greater detail. It is merely observed that the Notice contains provisions on the following topics:

(a) lead and cadmium in ceramic materials;[41]

(b) foods;

(c) tetraethyl and tetramethyl lead;

(d) engine coolants, etc.;

(e) petrol;

(f) aerosol containers;

(g) fire-extinguishing equipment in regard to content of halogen-substituted hydrocarbons;

(h) killing of fur-bearing animals with cyanide compounds and strychnine;

(i) insect-catching;

(j) paints, glues and adhesives;

(k) disinfestation agents and preservatives for masonry, woodwork, textiles, clothing and articles in everyday use;

(l) flame retardants in textiles;

(m) formaldehyde in chipboards;

PRODUCT CONTROLS

(n) decorative objects;

(o) goods for use in connection with hobbies, for domestic use and the like.[41]

10.5 OTHER REGULATIONS PURSUANT TO THE CHEMICALS LEGISLATION

This heading introduces a very brief discussion of other product regulations issued administratively by means of separate Notices in pursuance of the 'chemicals legislation' meaning the rules in the previous legislation which was superseded as of 1 October 1980 by the Chemicals Act, cf. especially sections 30 and 31 of the latter.

The provisions concerning *cosmetics* were originally included as a single subsection in the 1977 Notice referred to under 10.4. In Notice No. 38 of 2 January 1979 this subject was singled out for special and more comprehensive regulation in accordance with EEC Council Directive 1976/768. The regulations include, in particular, provisions relating to the composition, labelling and control of cosmetics.

In 1975 a Notice concerning the *degradability of washing and cleaning preparations* was enacted.[43] Under the central provision in section 2, washing and cleaning preparations containing surfactants with a mean biodegradability of less than 90% for each of various more precisely defined syndets must not be imported or used. If a preparation contains anionic active syndets of which the degradability is less than 80%, the Environment Agency must issue a ban on the importation and use of the washing or cleaning preparation, cf. section 6. The Notice also lays down rules concerning packaging and control methods which provide for the use of control methods approved by the EEC Commission.

In conjunction with EEC Council Directive 1976/403, Notice regulations concerning restrictions on the importation and use of *PCB and PCT* were implemented.[44] The Notice contains a more precise enumeration of the purposes for which PCB and PCT may each be used, some labelling rules and some regulations concerning control, including a reporting obligation for the producer and importer.

Finally, it should be mentioned that by means of a 1977 Notice regulations were issued concerning the *restriction of dioxins in pentachlorophenol, etc..*[45]

10.6 LEAD SHOT

Under section 6 of the Environmental Protection Act the Environment Minister issued, in Notice No. 8 of 9 January 1981, regulations concerning the use of lead shot for shooting at targets in flight. The Notice contains a general ban on shooting with lead shot and also shot containing lead or alloys thereof from areas where clay-pigeon shooting with shotguns takes place regularly with shot coming down over random shot areas. The shot range area is determined as a section of a circle with the centre at the stand site and a radius of 250 m. The county council (in the metropolitan area, the metropolitan council) may nevertheless waive this ban when the shot comes down over minor, well-defined random shot areas that are not important as wildfowl habitats, feeding areas and also when special circumstances ensure that there is no risk to these birds of lead shot poisoning.

10.7 OIL PRODUCTS

In regard to the sulphur content of fuel oil, etc. and the lead content of petrol, readers are referred to the exposition above under 2.1.2.2 and 2.2.4.

10.8 PACKAGING

Readers are referred to the account above under 7.7 concerning the use of non-returnable containers for beer and minerals, and also of the opportunities for regulation afforded by the Recycling Act.

10.9 RADIOACTIVE SUBSTANCES

In this connection, refer to the discussion of the subject under 9.2.

Notes

1. Acts Nos. 119 and 118 of 3 May 1961, respectively, as subsequently amended.
2. Cf. *Folketingstidende* 1978–79, annex A, col. 1333 ff, and Jens Steensberg, *Gifte og*

PRODUCT CONTROLS

      sundhedsfarlige stoffer ('Poisons and substances dangerous to health'), 1981, pp. 15–18.
3. Cf. *Folketingstidende, op.cit.*, col. 1353.
4. Cf. 10.6 below.
5. Cf. the whole of section 3 of the Act.
6. Section 5 of the Act.
7. Section 6 of the Act.
8. Cf. in this connection the somewhat parallel weighing-up provision in section 1, subsection 3, of the Environmental Protection Act, 2.1.2 above.
9. *Folketingstidende, op.cit.*, col. 1386.
10. Notice No. 409 of 17 September 1980.
11. Cf. 7.6.2 on the matter.
12. Cf. *Guidelines*, p. 9.
13. Cf. section 12 of the Notice and Guidelines, p. 18.
14. Cf. Working Environment Act, section 49a, subsection 4, and Guidelines, p. 18.
15. Cf. *Guidelines* p. 22.
16. Cf. 10.5 below.
17. Cf. 10.5 below.
18. Cf. sections 45–47 of the Act.
19. See regarding information and investigation obligation also sections 10, 16, 19, 34, 42, 43 and 49–51 of the Act.
20a. Cf. for further details Environment Agency Guidelines No. 2/1980, p. 12 *et seq.*
20b. This occurred to a limited extent in Notice No. 185 of 15 April 1981 concerning aerial spraying of pesticides, section 13, subsection 1.
21. Cf. for further details sections 59–63 of the Act.
22. In this connection see the first edition of the present publication under 9.1.
23. Cf. Environment Agency Guidelines No. 2/1981, p. 6.
24. Cf. *Folketingstidende, op.cit.*, col. 1402, and *Guidelines*, p. 7 *et seq.*
25. Cf. for further details, *Guidelines*, p. 30.
26. Cf. for further details, *Guidelines*, pp. 31–35.
27. Concerning the sales concept of the Act, see section 4 thereof.
28. Cf. for more details, *Guidelines*, pp. 36–38.
29. Cf. for further details, *Guidelines*, pp. 39 and 40.
30. Cf. for further details, *Guidelines*, p. 5 with references.
31. Act No. 118 of 3 May 1961 with subsequent amendments.
32. Cf. for more details, *Guidelines*, p. 6.
33. Cf. section 33 of the Act and section 2 of the Notice.
34. Sections 16–18 of the Notice.
35. Cf. Claus Haagen Jensen, in *Forvaltningsret, Almindelige emner*, 1979, p. 352 ff.
36. Readers are referred to the Notice together with *Guidelines*, pp. 9–11.
37. Cf. section 16, subsection 2, and section 18, subsection 2, of the Notice.
38. Cf. sections 24 and 25 of the Notice.
39. Notice No. 603 of 28 November 1973, as amended by Notice No. 341 of 17 June 1976.
40. Notices No. 38 of 2 January 1979, No. 224 of 22 May 1979 and No. 106 of 21 March 1980.
41. In this connection, see the Environment Agency report of October 1980 concerning the use, storage and harmful effects of cadmium in Denmark.
42. In regard to the details, readers may be referred chiefly to Jens Steensberg, *op.cit.*, pp. 223–312.
43. Notice No. 347 of 27 June 1975.
44. Notice No. 18 of 15 January 1976, which was implemented by Notice No. 572 of 26 November 1976.
45. Notice No. 582 of 28 November 1977.

11
Environmental Impact Assessment

Under the American National Environmental Policy Act (NEPA) of 1969, the federal administration was directed to prepare an Environmental Impact Statement (EIS) in conjunction with any proposal on 'major federal actions significantly affecting the quality of the human environment'. Subsequently, similar schemes for different variations on 'environmental impact statements' were considered in a number of West European countries and in some cases the debate has resulted in legislation on the subject.

Taking the American EIS regulations as a basis, but without being bound too closely by them, it seems reasonable to sum up the characteristic features of a scheme for environmental impact statements as follows:

(a) Before the public authorities reach a decision concerning projects significantly affecting the physical environment, a statement must be drawn up on the project and its anticipated effects.

(b) The statement should elucidate all types of environmental consequences of the project and assess these in the light of social values other than the environment.

(c) The statement is not confined to dealing with the project in isolation but also contains an assessment of it in relation to alternative measures with the same or similar objectives.

(d) The statement is designed to be an important element in the authorities' decision and also to serve as a basis for public discussion, affording all groups and individuals concerned an opportunity to come forward with their views and thus broaden the grounds for decision. The latter purpose is evident in the requirement that the statement be published, in provisions imposing a duty to draft the statement so that it is easy to read and in directing the authorities to initiate or stimulate public discussion by various means.

ENVIRONMENTAL IMPACT ASSESSMENT

The current Danish regulations do not contain any stipulations about preparing environmental impact statements which cover the four points set out. On the other hand, in at least one and possibly two sectors, arrangements exist which contain important elements of these points. This is true firstly within the general planning legislation—in this connection, see 2.1.1 above. Under the National and Regional Planning Act of 1973 and the Municipal Plan Act of 1975, procedures were introduced relating chiefly to regional and municipal plans and, to a lesser extent, to local plans to ensure that these plans could not be approved until a fairly comprehensive dossier had been compiled and the general public had been involved in the discussion of it. For the regional and municipal plans, the dossier must contain a broad survey of various planning alternatives and must be drawn up in such a way as to be readily understandable. The authorities have a duty not merely to publish the dossier but also to institute the public debate in various ways. These arrangements differ from the environmental impact statements chiefly in that the procedure concerns all forms of activities within the scope of the plan and is not, as for the environmental impact statements, centred on a single project.

Secondly, we can probably point to a similar procedure in the siting of nuclear reactor plants outlined in the Act on Safety and Environmental protection measures in atomic plants, etc. section 14, subsection 7; in this connection see 9.1.7.1 above. This procedure relates, like the typical environmental impact statement, to a particular finite project and involves the public before a decision is reached. On the other hand, the statutory regulations concerning reactor plants do not expressly stipulate that written material should be made public though it is clearly anticipated in the preparatroy works that a dossier would be prepared in conjunction with the public meeting to form a basis for discussion[1] and that the general public would be acquainted with the application dossier, including the safety document prepared by the applicant.[2] If these intentions set out in the preparatory works are followed up in due course, the arrangement relating to reactor plants will differ from an environmental impact statement chiefly in that there is no requirement to set out alternatives to the project, as there is with the latter.

The Danish authorities have not actually decided whether or not to introduce the preparation of environmental impact statements legislatively (on a wider scale) and, so far as is known, no considerations on this subject have been committed to print.[3] Yet from the fact that proposals for environmental impact statements have not been presented and no more light has been shed on the matter in official documents, it may be concluded that the administrative authorities, have no desire to see these introduced in Denmark, at any rate for the present.

Any discussion of the introduction of such requirements in Danish legislation must presumably be linked primarily with the planning legislation passed in the 1970s.[4] It could be asserted, in the first instance, that this legislation rendered environmental impact statements superfluous. No major project with significant environmental consequences can be executed without a plan which presupposes the preparation of a more or less comprehensive dossier and the involvement of the general public. This is, of course, correct in that the new planning legislation reduced the need for environmental impact statements. Yet there is scarcely any justification for urging that such statements are unnecessary. The considerations on which regional and municipal plans are based relate broadly to the future use of large areas and do not allow for the individually significant project to be thoroughly examined in an environmental context. In conjunction with local plans a single major project may well be of central interest but, on the one hand, the overriding pattern into which the local plan must fit is already determined by regional and municipal plans and, on the other, both the investigation and publication procedure are relatively restricted in the case of local plans. The scheme outlined in the 1976 Act on Nuclear Reactor Plants appears to confirm the need for a form of environmental impact statement to be prepared in really special cases.

Against this, it might be asserted that the introduction of environmental impact statements would destroy important sections of the new planning system. The main idea behind this is that planning must be based on coherent considerations regarding the utilisation of large areas. If environmental impact statements are made compulsory, a tendency might arise for individual cases to determine the planning of large areas to too great an extent and for too much priority to be given to environmental considerations. On this point it may be said, in particular, that the objection would indeed carry weight if environmental impact statements had to be prepared in a large number of cases but not if the scheme is limited to projects of exceptional importance such as, for instance, motorways, large bridges, oil refineries and major power stations. Decisions concerning such installations will probably always have an in-built tendency against being severed from the general planning. The difference could be that environmental impact statements would ensure a reasonably thorough clarification and public discussion of the environmental issues where today investigations are more restricted and there is often no discussion at all.

Notes

1. Cf. *Folketingstidende* 1975–76, appendix B, cols. 1032 and 3.
2. Cf. *Folketingstidende, op.cit.*, col. 1038, compared with col. 1020 ff.

3. A brief account of the American regulations is to be found in Bent Christensen, *TfR*, 1976, pp. 31–33 and 43.
4. This is not to overlook objections concerning the considerable resources required to prepare environmental impact statements which are especially hard to find at a time when efforts are being made to cut down in the public sector.

12
Closing Remarks

No attempt will be made to produce an overall conclusion relative to the preceding Chapters. This is not called for as the chief object of the account is to present a survey in fairly broad terms of the regulations that apply within the sector of 'environmental protection' with its many ramifications. However, it may be reasonable to make a few general comments on the system of environmental protection regulations as a whole, as it may appear to a jurist.

1. *Is there adequate statutory authority for intervention?*

From a practical viewpoint, this question is undoubtedly the most interesting in this concluding section. Is the content of the relevant statutory rules such as to afford satisfactory provision for effective action in regard to the protection of the environment? This question is at the same time the hardest for a jurist to answer because he is not professionally qualified to assess all the instances in which legal controls are needed on ecological, biological, medical and similar grounds.

With this reservation, it is my impression that the current Danish legislation contains, by and large, adequate provision for imposing legal requirements to protect the external environment. It should be stressed in this connection that a number of the weaknesses associated with this aspect of the state of law immediately after the Environmental Protection Act took effect on 1 October 1974 have been remedied by new legislation. This includes, in particular, the Marine Environment Act 1980 (Chapters 4 and 5), the Recycling Act 1978 (Chapter 7) and the Chemicals Act 1979 (Chapter 10). During my survey I have not come across any fields where there appears to be an obvious need for new powers of control. It may perhaps nevertheless be put forward that the creation of wider authority to issue binding general requirements for the quality of the environment seems to be called for. Binding quality regulations can, to be sure, be incorporated in the regional plans, but it is doubtful whether this would suffice to meet the need for quality targets resulting *inter alia* from Danish participation in Common Market collaboration. It also seems invidious that the approval regulations of the Environmental Protection Act did not incorporate powers to revise the terms of an approval after a period of, say, 10 years; see 2.1.2 above.

This general finding that adequate statutory authority exists cannot be taken to mean that there are sufficient 'operative' general regulations. In numerous respects the authority to intervene is provided in the form of powers for the Environment Minister especially to issue regulations and many of these powers have so far been utilised only to a limited extent. Whether this will be done on a broader scale than heretofore is not a question that a jurist is specially qualified to answer.

Indeed, the answer depends mainly on the scientific knowledge available and the resources in terms of money and manpower which society has at its disposal and is willing to invest.

2. *The administrative organisation*

On the question of administrative organisation, the 1973 environmental reform involved sweeping changes aimed at remedying the manifest defects in the earlier legislation. In particular, there was considerable expansion of the central administrative apparatus in the environmental protection sector and the local tasks were predominantly entrusted to the local councils, county and municipal authorities. With certain minor adjustments, this system has been retained over the intervening period.

My impression is that this administrative organisation functions quite well and it is scarcely possible to point to another system as unquestionably better. In this context it must be realised, among other things, that we cannot just assess the presumed efficiency of the organisation in terms of environmental protection but must also take account of how it fits into the administrative structure in general. The far-reaching 'decentralisation' of local administrative tasks in numerous sectors in the 1970s is in itself an important argument for giving the local councils considerable powers relating to environmental protection.

The scepticism that was expressed chiefly in the first few years after the 1973 environmental reform appears to have diminished. The counties and municipalities have, in general, proved equal to dealing with the environmental protection tasks assigned to them. The question arises, however, as to whether it is appropriate for decisions on approval under Chapter 5 of the Environmental Protection Act to be taken in virtually every case by the municipal councils. In a number of especially complex—and typically 'big'—cases, these councils do not have the professional expertise at their disposal to be able to reach a decision on their own. Consequently proposals to amend the law are at present being prepared at the Environment Ministry, in accordance with which the powers of approval are transferred to the Environment Agency in certain categories of cases. There has also been some criticism of the composition of the Environmental Appeal Board, which may be said to be biased because the 'producer side' is represented whereas the 'consumer side' is not, cf. 1.3 above. However, the decisions by the Appeal Board on record do not, in my opinion, indicate that this fact has had repercussions in practice, on the whole. This does not mean to say that changes in the make-up of the Board would be unreasonable and it is understood that proposals to this effect are under consideration at the Environment Ministry.

3. The individual's legal security

Legal security may be briefly expressed as a question of whether the individual has reasonable opportunities of adjusting to the legal requirements and of having his views thoroughly and objectively examined. The answer to this question depends mainly on (a) the degree of precision of the material regulations, (b) the hearing rules, and (c) the facilities for control of administrative decisions.

In regard to the *precision of the material regulations*, the fundamental point under the Environmental Protection Act is that the decisions should be reached after an assessment. To this extent, the legal security is extremely slight. However, the bases of decision have been very considerably standardised through the Environment Agency issuing a large number of recommendations. To be sure, these are not binding but the citizen can be absolutely certain that if these are adhered to, no further requirements will be imposed and that he will usually be required to comply with the standards in the guidelines. By this means legal security has been substantially safeguarded without losing the flexibility of the system. More precise rules could undoubtedly be introduced but, in my view, the assessment cannot be eliminated unless we are willing to accept unfair results in the form of overprotection of the environment or environmental demands that are too lenient.

The administrative procedure in environmental cases appears to be satisfactory by and large. The individual—polluter as well as person exposed to pollution—has ample opportunity of obtaining information and of making his views known in the proceedings; cf. 1.7 above, in particular. In this context the current arrangements are to my mind nevertheless defective in one important respect. During the initial handling of an environmental case, persons other than the polluter will often be unaware of its existence because there is no general obligation to notify 'third parties'. This weakness should be remedied. A limited and not entirely satisfactory solution would be, similar to the American pattern, to introduce the obligation to prepare environmental reports with the associated publication procedure in cases of especially wide significance from an environmental viewpoint. Such an arrangement would also, and perhaps primarily, have advantages other than those relating to legal security; for further details, see Chapter 11.

The *control* of administrative environmental protection decisions is generally excellent from the point of view of legal security. Control by the courts is certainly more a matter of theory than practice but the administrative appeal facilities have been developed in such a way that the individual can have virtually all decisions examined by one superior administrative instance and most decisions of major importance by two, the last one being an independent appeal board. Two criticisms may nevertheless be levelled at the control system, only one of which concerns legal security.

First, it is doubtful whether the regulations concerning the *right of appeal* are satisfactory. It is true that a third party who has a substantial individual interest in the outcome of the case has a right of appeal on a par with the polluter, cf. section 74 of the Environmental Protection Act. Yet this provision has been interpreted in practice in such a way that organisations can almost never appeal, unless acting on behalf of an individual entitled to appeal, and the interest of these individuals is simultaneously assessed primarily on the basis of their geographical proximity to the source of pollution – a form of 'extended adjoining property criterion'. The result is that the various environmental organisations and associations often have difficulty in getting their grievances heard and that in certain types of case—especially those concerning the discharge of waste water into the sea—no-one is entitled to appeal other than the polluter and some specially listed authorities; in this connection, see 1.7.3 above. This is not satisfactory, nor does it contribute to the efficiency of environmental protection. A revised version of the Act being prepared in the Environment Ministry actually incorporates a proposal for broader rights of appeal for organisations.

Second, it may be objected that the present control system is *too ponderous* or, to put it another way, that in this respect legal security is 'overprotected'. Such a criticism may chiefly be directed at the facilities for bringing too many cases before two appeal authorities, in the last instance the Environmental Appeal Board. The fact that Chapter V of the Environmental Protection Act relating to approval, etc. of specially polluting enterprises covers a great many types of installation and that the appeal against all decisions under Chapter V falls to the Environment Agency and from there to the Appeal Board results in the latter dealing with many cases of quite modest proportions. It is reasonable enough that the Appeal Board should be able to deal with 'major' environmental cases and also those relating to questions of principle and of real importance to the person or persons concerned. On the other hand, it seems doubtful whether it is reasonable for so much of the Board's time to be taken up with minor cases concerning noise and smell (for instance, from a mink farm) as happens today.

4. Enforcement of environmental requirements

For legal controls in a sector to be described as satisfactory from a juridical viewpoint, enforcement must meet an impressive target of effectiveness. If the requirements merely exist in the texts of Acts and Notices and specific administrative decisions without being respected or enforced, the system is imperfect in the extreme. This is probably the weakest point of the present system in the sphere of environmental protection.

This fault is not due to the fact that no-one has attempted to tackle the problem. The transition in conjunction with the Environmental Protection Act to the widespread use of permit schemes as a means of control rather than orders or prohibitions has, of course, benefited the enforcement authorities. The right of private individuals other than the polluter to appeal against administrative decisions has been useful in enlisting private persons in the business of supervision, even though to an inadequate extent, cf. Chapter 3. Far greater resources than before the 1973 environmental reform have also been expended on supervision by the authorities—cf. especially 4.1.7 above—and efforts have been made to get the courts to react more rigorously in cases of punishable violations of environmental requirements, cf. 2.1.6.2 above.

Nevertheless, enforcement and hence compliance is, in the environmental protection sector as in many other areas of statutory controls, relatively poor. The weaknesses emerge in two ways: first, the rules discussed in the preceding chapters are violated as a matter of course in a very large number of cases although naturally there are no further details of the extent of the violations becuase they do not actually come to the authorities' knowledge. The supervision is quite inadequate in many fields. This applies not least to the regulations concerning marine pollution (cf. Chapters 4 and 5) but in other sectors too, e.g. in regard to the requirements for motor vehicles (2.2 and 8.2) and air pollution not due to dust, smoke or smell, many infringements pass unnoticed.

Second, the authorities' reactions to violations that are actually discovered are often too weak, so that the incentive to comply 'voluntarily' is undermined. On the one hand, only relatively low penalties are still being imposed for infringements of environmental protection legislation (cf. 2.1.6.2 above) and a cynical works management may be tempted, for instance, not to obtain the necessary permission before starting an operation simply because the fine is less than the financial profit from an early start to the operation. On the other hand, the municipal super-

CLOSING REMARKS

visory authorities do not in some cases react strongly enough to infringements, by using *inter alia* administrative means of coercion. Yet instances of this occur more frequently relative to the building and planning legislation than the environmental protection regulations.

Both these types of deficiency can be substantially put right. Supervision can be improved by extending the rights of the environmental organisations to take up relevant cases, with the appeal authorities too (cf. Chapter 3) thereby exploiting a widespread popular interest in a clean environment. Supervision by the authorities can also be increased and made more effective *inter alia* by making testing easier, cf. 8.2.5 above. It should also be possible to tighten up court practice in the matter of imposing penalties even though this has so far proved difficult to do. The possibilities of introducing greater responsibility on the part of mayors and other municipal council members are at present being considered by a committee set up by the Interior Ministry.

Although improvements in enforcement are feasible, we must realise that a perfect or even anything like perfect system of enforcement lies beyond the bounds of possibility, nor would it be really desirable. The enforcement of regulations is not only a problem in the environmental protection area. It is a fundamental and general issue in the increasingly 'regulated' life of modern society. The limits for its solution within environmental protection are fixed not only by the practical, economic and political resources available in this sector but also by the ability and will of society and the individual to solve the problem of the enforcement of legislative controls in general.

Classified Index*

The Constitution, Public Authorities, Special Interest Groups and Individuals

The national constitution	1.1
Sources of laws governing pollution control and remedies for damage caused by pollution	1.2
Government departments and agencies with supervisory, administrative or executive powers of pollution control	1.3
National, regional and local public authorities with powers of pollution control	1.3, 1.4
Independent advisory bodies with rights or duties under pollution control legislation	1.5
Special interest groups representing those who may be liable for pollution, or those concerned to prevent or reduce pollution	1.6
Standing to sue *(locus standi)* in legal proceedings for pollution	1.6, 1.7.3

Air

Stationary Sources   2.1

Control by land use planning	2.1.1
Controls over plant and processes (including raw materials, e.g. fuels)	2.1.2
Controls over treatment before discharge, and over manner of discharge (e.g. height of chimney)	2.1.3
Limits on emissions	2.1.4
Monitoring to be done by discharger	2.1.5
Enforcement, including monitoring and surveillance by or on behalf of the control authority	2.1.6
Ambient air quality standards	2.1.7
legal requirements, objectives and guidelines	2.1.7.1
evidence of achievement of standards	2.1.7.2

* References are to section numbers.

CLASSIFIED INDEX

Rights of the individual	2.1.8
to be informed	2.1.8.1
to appeal against any permission to discharge	2.1.8.2
to have complaints or representations taken into account	2.1.8.3
to initiate or intervene in the enforcement process	2.1.8.4
to compensation and injunction	2.1.8.5
Road Vehicles	2.2
Controls over design and construction	2.2.1
Requirements concerning maintenance	2.2.2
Controls over use, including time and place of use	2.2.3
Fuel which may be used or supplied for use	2.2.4
Enforcement, including monitoring and surveillance by or on behalf of the control authority	2.2.5
Rights of the individual	2.2.6
Aircraft, Hovercraft and Ships	2.3
Aircraft	2.3.1
Hovercraft	2.3.2
Ships and other waterborne vessels	2.3.2, 2.3.3

Inland Waters (Including Groundwaters)

Stationary Sources	3.1
Control by land use planning	3.1.1
Controls over processes, storage and other operations near inland waters (including waste disposal)	3.1.2
Controls over treatment before discharge and treatment plant	3.1.3
Controls over characteristics, quantities and rates of discharge	3.1.4
Controls over dumping in and on land near inland waters	3.1.5
Monitoring to be done by the discharger	3.1.6
Enforcement, including monitoring and surveillance by or on behalf of the control authority	3.1.7
Inland water quality standards	3.1.8
legal requirements, objectives or guidelines	3.1.8.1
evidence of achievement of standards (may be by reference to sources of information only)	3.1.8.2
restrictions on use of waters not meeting standards	3.1.8.3
Rights of the individual	3.1.9
Mobile Sources	
Waterborne craft	3.2

CLASSIFIED INDEX

Seas: Pollution by Substances other than Oil   4

Coastal Waters   4.1

Definition of boundaries	4.1.1
Control by land use planning	4.1.2
Controls over processes, storage and other operations near coastal waters (including waste disposal)	4.1.3
Controls over treatment before discharge	4.1.4
Controls over characteristics, quantities and rates of discharge	4.1.5
Monitoring to be done by discharger	4.1.6
Enforcement, including monitoring and surveillance by or on behalf of the control authority	4.1.7
Coastal water quality standards	4.1.8
legal requirements, objectives or guidelines	4.1.8.1
evidence of achievement of standards (may be by reference to sources of information only)	4.1.8.2
restrictions on use of coastal waters not meeting standards	4.1.8.3
Rights of the individual	4.1.9

Controls over Dumping from Ships and Aircraft   4.2

Geographical extent of jurisdiction	4.2.1
Monitoring	4.2.2
Enforcement	4.2.3

Control of Pollution Caused by Exploitation of the Sea and Sea Bed   4.3

Geographical extent of jurisdiction	4.3.1
Monitoring	4.3.2
Enforcement	4.3.3

Seas: Pollution by Oil   5

Controls over Discharges from Ships   5.1

Construction and equipment	5.1.1
Crew	5.1.2
Discharges of oil	5.1.3
Facilities at ports for oily residues	5.1.4
Enforcement, including surveillance by or on behalf of the control authority	5.1.5
Civil liabilities	5.1.6

Controls over Shore Installations and Port Activities   5.2

Construction of shore installations	5.2.1

CLASSIFIED INDEX

Loading and unloading operations, including bunkering — 5.2.2
Enforcement, including surveillance by or on behalf of the control authority — 5.2.3

Controls over Offshore Installations and Operations — 5.3

Geographical extent of jurisdiction — 5.3.1
Licensing — 5.3.2
Construction, equipment, safety zones — 5.3.3
Manning — 5.3.4
Discharges — 5.3.5
Loading and unloading — 5.3.6

Contingency Plans for Oil Pollution Incidents — 5.4

Discharges to Sewers — 6

Prohibitions — 6.1
Controls over characteristics, quantities and rates of discharge to sewers — 6.2
Requirements for treatment of collected sewage — 6.3
Controls over quality of discharges from sewers or sewage treatment plants — 6.4
Enforcement, including monitoring and surveillance by the control authority — 6.5
Civil liabilities — 6.6

Disposal of Waste on Land — 7

Controls by land use planning, including licensing of sites for treatment and disposal — 7.1
Controls over treatment before disposal, including recycling and reclamation — 7.2
Controls over methods of disposal — 7.3
Restoration of land after tipping — 7.4
Enforcement, including surveillance by or on behalf of the control authority — 7.5
Controls over disposal of specified wastes, e.g. waste oils, old vehicles, wastes from particular industries — 7.6
Controls over specified classes of products, e.g. beverage containers, for purposes of reducing waste or aiding resource recovery — 7.7
Rights of the individual — 7.8

Noise and Vibration — 8

Stationary Sources — 8.1

CLASSIFIED INDEX

Controls by land use planning, including licensing	8.1.1
Controls over design and construction of noise generating plant and equipment	8.1.2
Restrictions on emissions	8.1.3
Enforcement, including monitoring and surveillance by or on behalf of the control authority	8.1.4
Ambient noise standards	8.1.5
legal requirements, objectives or guidelines	8.1.5.1
evidence of achievement of standards (may be by reference to sources of information only)	8.1.5.2
restriction of activities in noisy areas	8.1.5.3
requirements for insulation	8.1.5.4
Rights of the individual	8.1.6

Road Vehicles   8.2

Controls over design and construction	8.2.1
Emission standards	8.2.2
Requirements as to maintenance	8.2.3
Controls over use, including times and places of use	8.2.4
Enforcement, including surveillance by or on behalf of the control authority	8.2.5
Rights of the individual	8.2.6

Aircraft   8.3

Controls over design and construction	8.3.1
Emission standards	8.3.2
Requirements as to maintenance	8.3.3
Controls over use, including times and places of use	8.3.4
Enforcement, including surveillance by or on behalf of the control authority	8.3.5
Rights of the individual	8.3.6

Other Mobile Sources   8.4

Nuclear Energy   9

Nuclear Installations   9.1

Controls by land use planning, including licensing	9.1.1
Controls over design and construction	9.1.2
Controls over maintenance and operation	9.1.3
Controls over accumulation of discharge of wastes	9.1.4
Monitoring to be done by discharger	9.1.5
Enforcement, including monitoring and surveillance by or on behalf of the control authority	9.1.6
Rights of the individual	9.1.7

CLASSIFIED INDEX

Radioactive Substances — 9.2

Controls by land use planning, including licensing	9.2.1
Controls over storage and use	9.2.2
Controls over packaging and transport	9.2.3
Controls over accumulation and disposal of wastes	9.2.4
Monitoring to be done by the discharger	9.2.5
Legal standards, objectives or guidelines for levels of radioactivity in the environment	9.2.2, 9.2.6
Enforcement, including monitoring and surveillance by or on behalf of the control authority	9.2.7
Rights of the individual	9.2.8

Controls over Products[i] — 10

Pesticides	10.3
Veterinary products	10.1
Detergents	10.5
Fuels	10.7
Packaging	10.8

Environmental Impact Assessment

Requirements for environmental impact assessment	11

---

[i] I.e controls over products introduced for the purpose of protecting the external environment.